Social Policy Reform
IN **HONG KONG** AND **SHANGHAI**

HONG KONG BECOMING CHINA

Ming K. Chan and Gerard A. Postiglione
Series General Editors

Ming K. Chan is Research Fellow, and Executive Coordinator of the Hong Kong Documentary Archives, Hoover Institution, Stanford University.

Gerard A. Postiglione is Associate Professor of Education, University of Hong Kong.

In view of Hong Kong's remarkable historical development under British rule, its contemporary global importance as an economic hub and communication center on the Pacific Rim, and because of its new status as a Special Administrative Region (SAR) of the People's Republic of China since July 1, 1997, M. E. Sharpe has launched this multivolume series for an international readership. This series aims at providing both expert analysis and the relevant documentary basis for an informed appreciation of the key issues and major dimensions of Hong Kong's developmental experience. The life and work, hope and fear of the seven million people in the Hong Kong SAR as a free society, capitalist economy and cosmopolitan community under socialist Chinese sovereignty should be of direct relevance to those interested in Chinese affairs. The "one country, two systems" formula that has guaranteed the HKSAR's high degree of autonomy is also the cornerstone policy of the PRC's cherished peaceful reunification with Taiwan. As such, Hong Kong's record of performance as a part of the PRC is of crucial significance to the prospects of Greater China and the Asia-Pacific region in the twenty-first century.

The books in this series:

THE HONG KONG BASIC LAW:
BLUEPRINT FOR "STABILITY AND
PROSPERITY UNDER CHINESE
SOVEREIGNTY?
Ming K. Chan and David Clark, editors

EDUCATION AND SOCIETY IN
HONG KONG: TOWARD ONE
COUNTRY AND TWO SYSTEMS
Gerard A. Postiglione, editor

THE COMMON LAW SYSTEM IN
CHINESE CONTEXT:
HONG KONG IN TRANSITION
Berry Hsu

PRECARIOUS BALANCE: HONG
KONG BETWEEN
CHINA AND BRITAIN, 1842–1992
Ming K. Chan, editor

RELUCTANT EXILES? MIGRATION
FROM HONG KONG AND
THE NEW OVERSEAS CHINESE
Ronald Skeldon, editor

THE HONG KONG-GUANGDONG
LINK: PARTNESHIP IN FLUX
*Reginald Yin-Wang Kwok and
Alvin Y. So, editors*

HONG KONG'S REUNION WITH
CHINA: THE GLOBAL
DIMENSIONS
*Gerard A. Postigline and
James T.H. Tang, editors*

CRISIS AND TRANSFORMATION IN
CHINA'S HONG KONG
Ming K. Chan and Alvin Y. So, editors

SOCIAL POLICY REFORM IN
HONG KONG AND SHANGHAI:
A TALE OF TWO CITIES
*Linda Wong, Lynn White, and
Gui Shixun, editors*

Social Policy Reform

IN **HONG KONG** AND **SHANGHAI**

A Tale of Two Cities

Linda Wong, Lynn White, and **Gui Shixun,** editors

An East Gate Book

M.E.Sharpe
Armonk, New York
London, England

An East Gate Book

Library of Congress Cataloging-in-Publication Data

Social policy reform in Hong Kong and Shangai: a tale of two cities / edited by Linda Wong,
Lynn T. White and Gui Shixun.
    p. cm.
    "An East Gate Book."
    Includes bibliographical references and index.
    ISBN 0-7656-1311-5 (cloth)
    1. Hong Kong (China)—Social policy. 2. Shanghai (China)—Social policy. I. Wong, Linda,
1949– II. White, Lynn T. III. Gui, Shixun.
HN752.5.S66 2003
361.6'1'0951132—dc21

2003008361

Printed in the United States of America

The paper used in this publication meets the minimum requirements of
American National Standard for Information Sciences
Permanence of Paper for Printed Library Materials,
ANSI Z 39.48-1984.

BM (c)   10   9   8   7   6   5   4   3   2   1

Contents

List of Tables and Figures

Tables

Figures

Series General Editors' Foreword

Historically, both Shanghai and Hong Kong share a common British colonial heritage. The British defeated Qing Dynasty China in the 1839–42 First Opium War. The end of the war led to the 1842 Treaty of Nanking, which ushered in China's "century of unequal treaties" humiliation under foreign imperialism. The Nanking treaty ceded Hong Kong island to the British crown and also opened up five "treaty ports" on the South China coast for international commerce and foreign settlement—Canton, Amoy, Foochow, Ningpo, and Shanghai. Hence, the history of modern Shanghai with its British-dominated International Settlement to some extent paralleled the growth of Hong Kong as a British colony since the mid-nineteenth century.

By the turn of the twentieth century, Shanghai had become the premier commercial, banking, and international shipping hub of the Chinese mainland, while Hong Kong developed into the second most important international port on the China coast as well as the key outpost for the British economic empire in the Far East. Indeed, until the Pacific War, Shanghai and Hong Kong were often regarded as sister cities under foreign imperialism—the twin international ports that symbolized Western economic penetration and sociocultural presence in coastal China. Through the entrenched British mercantile networks, the Cantonese entrepreneurs based in and working through colonial Hong Kong managed to establish themselves as the second most powerful Chinese economic bloc in Shanghai, right after the Ningpo natives hailing from Shanghai's southern reach. In pre-World War II Shanghai, the largest international banking house was none other than the local branch of the Hong Kong and Shanghai Banking Corporation (HSBC), which was incorporated and headquartered in Hong Kong. The "Big Four" department stores (Wing On, Sincere, The Sun, and Sun Sun) on Nanking Road in Shanghai's International Settlement were all Hong Kong/ Cantonese establishments. The single largest private Chinese industrial manufacturing plant then in Shanghai also happened to be another Hong Kong/Cantonese offshoot, the Nanyang Brothers Tobacco Company. Furthermore, most of the modern steamship traffic out of Shanghai harbor, both inland/coastal and overseas, was dominated by the major British concerns of Butterfield & Swire, Jardine & Matheson, and the P&O lines, which were all remote-controlled from their Chinese regional headquarters in colonial Hong Kong, where some of the ships were registered under the British flag and many of their crew were recruited. It was only natural that during the 1920s the national headquarters of the Chinese Seamen's Union operated from Hong Kong, its place of birth. Because of the longtime British commercial relationship with the

Cantonese, many of the top compradores in big British firms that dominated Shanghai's commercial landscape were Cantonese from Hong Kong.

The Hong Kong/Cantonese presence, often riding on British coattails, was so pervasive and influential in prewar Shanghai that all the mayors of Shanghai Municipality during the 1928–37 Nanking Decade of Kuomintang rule were Cantonese. In the early years of the 1937–45 Sino-Japanese War, a host of major Chinese enterprises and banks in Shanghai took refuge in colonial Hong Kong after the Chinese part of Shanghai fell into Japanese hands in late 1937. Some of these Shanghai firms (such as the Commercial Press) either set up sizable branches in the British enclave in the mid-1930s ahead of the Japanese invasion or undertook wholesale relocation of their key personnel and main operations to the colony soon after the war broke out in July 1937. Likewise, numerous Chinese public institutions (such as the National Committee of the Chinese Red Cross) and semiofficial agencies (such as the Bank of China) as well as many senior Chinese officials, who used to operate out of Shanghai, were evacuated to the still neutral British colony. Then, on December 8, 1941, the same day as the Pearl Harbor attack, the Japanese forces took over the British-dominated Shanghai International Settlement while launching a military assault against Hong Kong. The British colony fell to the Japanese on Christmas Day 1941 and endured three years and eight months of military rule. Once again, the two sister cities shared the same fate under foreign imperialism, this time Japanese occupation. The end of World War II in September 1945 did not bring easy recovery to either Shanghai or Hong Kong.

The 1946–49 Chinese Civil War and the Communist victory in October 1949 drove massive waves of mainland Chinese in an exodus into Hong Kong, whose population had swollen rapidly from about two-thirds of a million in late 1945 to over two million by 1950. Among the newcomers were many Shanghai entrepreneurs who came south to Hong Kong with their capital, skilled technicians, machinery, business know-how, and international commercial contacts. Such an influx of immigrant Shanghai economic players constituted vital new forces that made an indispensable contribution to postwar Hong Kong's industrialization and economic takeoff as well as its eventual elevation to the status of world-class service hub.

In some sense, the sophisticated commercialism, the urbanism, and the international cultural milieu, as well as the global business outreach of "old" semicolonial and multijurisdictional (the Chinese municipality, the International Settlement, and the French Concession) Shanghai, while not compatible with the Communist new order after 1949, had found a happy and fertile ground in capitalistic and cosmopolitan Hong Kong. Thus, it did seem that Communist Shanghai's loss was capitalist Hong Kong's gain. Such well-known Hong Kong names as Shaw in the movie industry, Pao and Tung in shipping, Woo and Tang in textiles all demonstrate the leading roles played by Shanghai business émigrés in the Hong Kong economic success story during the second half of the twentieth century. The Shanghai dimension has been so crucial to modern Hong Kong's transformation that descendents of these Shanghai emigrants are not only among the top local elites in many fields, two are currently at the apex of the Hong Kong Special Administrative Region (HKSAR) government— Chief Executive Chi-hwa Tung and Financial Secretary Henry Ying-yen Tang.

Following Hong Kong's July 1997 sovereignty retrocession to the People's Republic of China (PRC), the two former British colonial sister cities, Shanghai and Hong Kong, are finally reunited as integral parts of the Chinese national community. The mainland's reform and opening in the last quarter century has propelled Shanghai into the forefront of Chinese modernity and economic development. Contemporary Shanghai has more than regained its former prominence as China's flagship urban center and premier economic hub; it is now being hailed as the prime showcase for Chinese developmental vision, a success story that will be further amplified in the Shanghai World Expo of 2010. Of course, sizable investments and technical/managerial talents from Hong Kong have been a major and vital part of the positive external inputs buttressing contemporary Shanghai's spectacular rise. Some of their recent investments and new business in Shanghai represent a kind of "homecoming" operation for more than a few Hong Kong economic players whose families were originally Shanghai exiles who took shelter in the British colony because of the 1949 Communist victory on the mainland.

This remarkable reemergence of Shanghai as the new symbol of China's modernization and economic power came at a particularly low ebb in postcolonial Hong Kong's economic fortunes. In terms of territorial domain, population size, and total productivity, Shanghai far outshines Hong Kong. Thus, in the still ongoing crisis of self-doubt and lack of confidence in the HKSAR's uncertain prospect of economic restructuring and functional transformation, the image of a rising Shanghai has triggered a new fear in many Hong Kongers' hearts and minds that their once dynamic and versatile city of nearly seven million would soon be superseded by a vibrant Shanghai of seventeen million as the preeminent Chinese megametropolis and East Asian super hub on the cutting edge of the twenty-first century. In terms of its brand new infrastructure and ultra-modern "hardware" development (not least the Pudong airport magnetic levitation train), many recent visitors believe that Shanghai has indeed surpassed Hong Kong in scope, breadth, and range. There is an unfounded and unjustified, yet widespread, fear among some Hong Kong pessimists that Shanghai's miraculous rebirth will soon spell Hong Kong's doom.

This latest chapter of the Shanghai story clearly confirms its successful reclamation— after a thirty-year hiatus of Maoist transformation that saw growth mainly in size—of its former natural eminence as the foremost Chinese inland and ocean port city at the strategic juncture where the Yangzi meets the open sea. But this did not and should not necessarily come at Hong Kong's expense. Just as they did during their prewar complementary co-existence as the international economic nodal points along the China Coast, Shanghai and Hong Kong can serve as the twin powerhouses for the marketization and globalization of the continent-size Chinese mainland with its massive 1.3 billion population. As then PRC premier Zhu Rongji aptly observed during his spring 1999 visit to the United

States and Canada, Hong Kong and Shanghai are the two jet engines on each side that propel the plane that is China's economic transformation.*

In response to the competitive pressure and potential displacement effects of Shanghai at the apex of the larger, richer Yangzi Valley hinterland, Hong Kong has recently geared up its collaborative efforts with its Pearl River Delta (PRD) hinterland in a host of areas. Such HKSAR-PRD cooperative undertakings range from shared strategic vision and coordinated developmental approach to joint infrastructure projects (such as a Hong Kong-Zhuhai-Macau bridge) and practical cross-border functional integration in such areas as customs and immigration control. Under both private sponsorship and governmental auspices, there have been numerous research projects, professional forums, academic conferences, and publications on various aspects of the HKSAR-Guangdong interface as a hot topic for public discussion and business planning. In contrast, in-depth, detailed, balanced, and informed scholarly studies on Hong Kong and Shanghai from a genuinely comparative perspective are few, despite increasingly popular awareness of the significant impact that their interactive dynamics will have on the course of China's development. The present volume has done much to fill a large void of serious academic work in the English language on this subject.

Indeed, no two major cities in China today have the mutual admiration, intense competition, and shared destiny of Hong Kong and Shanghai. While Shanghai has the political advantage of intimate ties to the national leadership and the natural geostrategic advantage of its Yangzi hinterland, it craves the high degree of autonomy, extensive external networks, capitalistic expertise, global orientation, and advanced legal-administrative "software" enjoyed by Hong Kong. Hong Kong knows that the past decades of its prosperity would not have been possible without imported Shanghai talents. At the same time, both realize that they are bound, whether by the forces of globalization or through Chinese national sovereignty, to gradually convergent paths in the twenty-first century. This is already evident in their linkages in economic development and popular culture, in trade and tourism, and in migration and manpower. Yet, the picture is less clear in the realm of social policy. For example, although their universities sometimes share staff and students and also collaborate in exchange programs and joint projects, their educational systems tend to socialize and produce different kinds of national citizen.

To be successful, the design and execution of social policy must connect intimately with the traditions and experiences that have united and shaped communities. Here, the experiences of these two metropolitan communities over the past half

---

*The above sketch of the Hong Kong–Shanghai historical links is based in part on three of Ming Chan's recent academic lectures: (1) "The Hong Kong/Guangdong Cantonese Economic Empire in Shanghai, 1842–1941" at the University of Queensland, July 2000; (2) "Hong Kong Between Shanghai and Singapore" at the Wharton School of Business, University of Pennsylvania, November 2002; and (3) "A Tale of Three Cities: Shanghai, Hong Kong, and Singapore" at Mount Holyoke College, April 2003.

century could not have been more divergent, as one struggled with Communist revolutions (and an ultraradical Cultural Revolution) while the other was a community of flight and resistance, of opportunity and markets. The social capital upon which daily life draws for action and commitment has been in transition in both places. For Shanghainese, the social relations of the *danwei* (work unit) are less important than the social relations of the company, market, and neighborhood district. For Hong Kongers, heavy reliance on overseas Chinese social capital is now being matched by the increasing importance of *neidi* (inland) social networks for business, housing, and even labor. Does this make the cities too different to compare?

All social science should be comparative, and any study of these two world-class Chinese cities must confront the obvious. No matter how much they differ in size, structure, capacity, and limitations, they look to the same science when it comes to finding the best way to medicate and house, educate and employ, protect and care for their citizens. Their leaders often join similar study tours to learn how other world cities experiment and operate in social-policy spheres. They measure their success against the global league tables in quality of life. Despite their apparently competitive relationship, Hong Kong and Shanghai share what they learn from and with one another as the dual preeminent Chinese sister cities.

In one sense, the once wide gulf between Hong Kong and Shanghai in social policy may be narrowing because of their latest developmental experience. On the one hand, the increasingly heavy interventionist hand and strong paternalistic instincts of the state in social welfare during the sunset of the British colonial era and in the initial years of the HKSAR have very substantially reshaped Hong Kong's social-policy approach during the last two decades. Laissez-faire capitalism with minimal social engineering is no longer an accurate description of Hong Kong's current economic- and social-policy orientation. The prolonged post-1997 economic recession and the resultant high unemployment as well as general impoverishment have created new demands on the state for direct governmental economic relief and more social assistance. The annual influx of some 54,000 legal immigrants from the Chinese mainland since the 1980s (that is, a daily quota of 150) has also created heavy pressure on social services and public resources in Hong Kong, especially during a period of serious budget deficits.

In Shangai, on the other hand, the fast-expanding private sector, which is driven by the market mechanism and profit incentives and assisted by external capital investments and managerial/technical talents, has yielded a new and growing middle class and a more fluid social scene via a diminishing state sector. The larger private economy and curtailed state may eventually influence the resources and policies for Shanghai's social development in the years to come. The extensive and rapidly accelerating urbanization trends in the mainland will definitely make the Shanghai experience in social-policy reform highly relevant to understanding the notions and realities of urbanism in China's national development in the coming decades. These are just some of the issues, observations, and prospects that will benefit from the careful analysis, informed delineation, and insightful comparisons afforded by this volume.

This volume represents the fruits of research of a team of twenty-two scholars (thirteen are from the City University of Hong Kong and seven are Shanghai-based) under the joint editorship of a Hong Kong social scientist and a Shanghai academic in collaboration with a leading American China scholar whose knowledge of Hong Kong and Shanghai is based on extensive firsthand experience in both cities. In their nine substantive chapters filled with reliable data and careful analyses, these Hong Kong/Shanghai specialists offer us a systematic delineation of the policies and plans as well as fair-minded assessments of the successes and failures of government inputs into social life in these two complex urban cores.

What have Hong Kong and Shanghai learned from each other? How empowered by social policies are their respective populations? Where are they along the global continuum? Is socialism as manifested in contemporary Shanghai still benevolent to its metropolitans and is the HKSAR style of capitalism still as callous as its reputation suggests? Are these two cities becoming as interdependent for labor, housing, and education as they are for finance and trade? In what ways should the citizens of these cities look to closer national sociocultural integration amid economic globalization to make their lives more rewarding, productive, and secure? Readers of this volume are invited to reflect on these very pertinent issues as the following pages reveal the contrasting yet overlapping tales of these two Chinese super cities that represent two probable developmental paths for twenty-first century urban China as a powerful economic hub on the Pacific Rim. As general editors of the *Hong Kong Becoming China* series, which was inaugurated in 1991, we are most delighted to present this splendid volume as our tenth title, one that is fittingly bi-focal by placing Hong Kong in a comparative light with its mainland counterpart, Shanghai.

Ming K. Chan and Gerard A. Postiglione
December 2003

Acknowledgments

We would like to thank the Center for Comparative Public Management and Social Policy (now renamed the Governance in Asia Research Centre) of City University of Hong Kong for the award of a Flagship Project Grant, which enabled us to complete this research and publication project. We would also like to thank City University's Department of Public and Social Administration for provision of clerical and administrative support in the preparation of the manuscript.

Social Policy Reform
IN **HONG KONG** AND **SHANGHAI**

1

Introduction

Linda Wong and Gui Shixun

Twin Sisters: Legacy and Transformation

Shanghai and Hong Kong are very much like blood sisters. Shanghai is called the Lustrous Pearl of the Orient; Hong Kong, the Pearl of the East. Like two stars on the China coast, the two cities have a lot in common, in terms of geography, history, and challenges. Strategically placed at the mouth of the Yangtze River and Pearl River estuaries, both cities command the physical and social capital of rich hinterlands. They also bear witness to the turbulent history of China as their destiny is linked to western forays into Qing China on her knees. The momentous stroke that rewrote their history is of course the Opium War and the Treaty of Nanking (1842). Foremost among the humiliations were territorial concessions. Hong Kong was ceded to Britain as a colony, a status that was to last until 1997. The same treaty also turned Shanghai into a treaty port and semicolony, when the principle of extra-territoriality took the physical form of the international settlements. However infamous the past may be, the insulation from the most vicious wars and turmoil that besieged the China heartland was the key to Shanghai and Hong Kong's success. In the process, the twins evolved into the great hubs of trade and industry unmatched by any other Chinese city. Of equal importance has been the absorption of hundreds of thousands of immigrants, among them entrepreneurs, intellectuals, revolutionaries, artisans, and ordinary people. The mixing of the most adventurous and enterprising elements from within China and from all over the world underlaid the fantastic blending of cultures and practices. This not only turned them into great metropolises, but it also gave a big push to China's modernization. At the same time, unbridled growth in the context of cultural bastardy sired many social problems—drugs, crime, prostitution, poverty, and exploitation—that bestowed both with an aura of exotic decadence.

Before the Second World War, Shanghai was undoubtedly the more developed of the two. By the 1930s, Shanghai had established herself as the most modern city in Asia, dubbed the Paris of the East. In the meantime, Hong Kong bloomed under British rule though never quite matching the glitter of her bigger sister. In the period between the two world wars, mutual fascination and friendly rivalry were key features of their relations, with economic and cultural exchange never slowing and both flourishing as oases from political turmoil.

The socialist revolution put a radical stop to the status quo. Not only this, communist success in 1949 spawned an exodus of Shanghai capital, skills, and people. What Shanghai lost became Hong Kong's gain. This influx speeded up Hong Kong's industrialization and maturation as a society. By the 1950s, Shanghai had been overtaken by Hong Kong in all aspects of social and economic development. On the other hand, transformation of Shanghai as a socialist city, along with communist aversion to cities as parasitic institutions, reversed Shanghai's fortunes. What's more, having the best-endowed economy, Shanghai was treated like the firstborn son in a patriarchal family. Given the burden of supporting parents and younger siblings, the bulk of local resources had to be surrendered. For example, between 1949 and 1983, Shanghai remitted 87 percent of her total revenue to the central government (Cheung, 1996, p. 55).[1] Such onus reduced Shanghai to a shadow of her former self. Politically and ideologically, the once freewheeling city came under the shackles of an authoritarian state. Civil society became suppressed, its much vaunted cosmopolitanism went under wraps to escape the wrath of deviance.[2]

Hong Kong, on the other hand, encountered a different fate. There may even be some truth to the saying that Hong Kong thrives on the woes of others. Examples of ironic gains are legion: influxes of money and talents from the Mainland, particularly Shanghai, after 1949 (Wong, 1988); big expansion in trade following China's embargo at the wake of the Korean War; boosts to tourism including "rest and relaxation" stints during the Vietnam War; asset inflows from the Chinese Diaspora in South East Asia; and so on. More generally, cheap imports of foodstuff, water, and industrial raw materials from the Mainland lowered business and living costs. This hidden subsidy was instrumental to the city's growth and prosperity. Politically Hong Kong carried on as a British colony. Despite its dependent status, the metropolitan country left the local administration pretty much alone. The government of Hong Kong also thought fit to intervene minimally in community life. In economic matters, the principle of laissez faire gave way to active nonintervention. This hands-off approach plus prudent financial stewardship enhanced Hong Kong's reputation as a bastion of free trade and ideal place to do business. Under the circumstances, it took Hong Kong less than thirty years after World War II to change from entrepot to industrial center. Ideologically, both government and people are committed to the values of free market, small government, and individualism.[3]

By the 1980s Hong Kong was ready for another transformation. The big chance came from the opening of China and the lure of its cheap land and labor. When the decade ended, most of Hong Kong's manufacturing industries had relocated north of the border. Deindustrialization gave way to a vast expansion of the financial, real estate, and tertiary sectors. This second rebirth propelled Hong Kong into an unprecedented period of growth and prosperity. By the middle of the 1990s, the per capita Gross Domestic Product (GDP) of Hong Kong had exceeded U.S.$22,000,[4] second only to Japan in Asia and overtaking Britain and some Western countries. Paradoxically such stunning growth coincided with a time of high tension as the colony's future came under negotiation by China and Britain. As a result public confidence took a battering in the time before the transfer of sovereignty.

In Shanghai, the onset of economic reforms in the late 1970s enthused but disappointed the city when it was bypassed for preferential treatment. In the ensuing decade, Shanghai enviously watched the emergence of Hong Kong as a global economic and financial center. It was not until 1990 when Beijing announced the opening of Pudong that Shanghai got the big break it was pining for. Thereafter, especially in the half decade before the new century, Shanghai's economic recovery was so stunning that sisterly rivalry revived. Ordinary Shanghainese openly aired their dream to rebuild their city's glory and to catch up with Hong Kong. Pride and optimism reached their zenith in October 2001 when Shanghai played host to the APEC Forum. International dignitaries in attendance were amazed at how fully it had shed its lackluster shell of a socialist city to become the most modern hub in China. Not surprisingly it was Hong Kong's turn to get the jitters. Two years into the new millennium, Hong Kong is still suffering from the aftermath of the Asian financial crisis. The prolonged recession is enough to generate dismay. For local residents, what is more disheartening is the loss of future direction, which completely saps their confidence. In relation to Shanghai, the fear of being surpassed becomes a near obsession. Of course the official line is quite different: both Hong Kong and Shanghai have their unique advantages, and they complement each other.

Social Policy Comparison: Why and What

The intertwining of historical and developmental destinies is perhaps the first starting point of our tale of two cities. In many respects, this rich mix of popular musings and introspection to their fate is an interesting phenomenon in itself. Such explorations have found their outlet in literature, film, music, and journalism (see, e.g., Lee, 2001; Wei, 1987). In the social sciences field, economics and political science have made their contribution to the twin-city comparisons (Wong, 1996; Cheung, 1996; Yao, 1990). As social policy researchers, we are interested in the study of the twins for other reasons. First, there is a vacuum of truly comparative scholarship on social development and social policy issues in these two cities.[5] It is our wish to plug this gap. We believe that the social chronicle of development is an important subject in its own right. Societies with abundant economic growth but little social development are lopsided societies. Their development may not be sustainable if the social fabric is allowed to fray, and worse, disintegrate through neglect and blind faith in trickle-down economics. We also believe that such a study is timely. We live in a restless, risk-prone, and even turbulent age. Globalization has quickened the pace of change and multiplied risks in a dramatic fashion. Both Hong Kong and Shanghai are undergoing rapid, wide-ranging, and deep transitions. A study on how the cities manage their developmental dilemmas, satisfy social needs, invest in human capital, and promote social integration is long overdue. The insight that derives from the comparison will promote self-reflection and mutual learning.

In their early beginning, Hong Kong and Shanghai were alike. When China turned communist, their trajectories diverged. The influence of their respective political, social, and economic systems continues to weigh heavily on state policies. This is as it should be. Notwithstanding such differences, both cities have to grapple with many common issues of the day. Indeed, these contemporary issues have universal salience to all big cities. How Hong Kong and Shanghai responded to these challenges is of immense interest to social scientists.

Before proceeding to the comparison, it is useful to mention some basic facts about the two cities. Table 1.1 summarizes the background information of the two cities in 2000. We have tried to include as much basic data across all social policy areas as possible, but some gaps are still unavoidable. To begin with, both have large populations. According to the 2000 census, Shanghai had a population of 16.7 million, and Hong Kong, 6.8 million. Such numbers make Shanghai one of the world's largest metropolises while Hong Kong also falls into the big-city league. Big populations spawn substantial needs. Among their minions, the weight of new immigrants is heavy in numbers and social impact. In 2000, Shanghai's floating population amounted to 3.9 million, a fivefold increase since the first sample survey on floaters was taken in 1984.[6] This means that one in four persons is a new arrival. In Hong Kong, between 1977 and 1996, some 800,000 to 900,000 immigrants from the Mainland had settled in the city. In the years since, 55,000 persons each year have been admitted for family reunions (Wong, 2001).[7] Thus within the last twenty-five years, Hong Kong has absorbed one million immigrants who comprise one in seven in the population. These demographic facts confirm their profiles as immigrant societies. The continuing inflow brings in valuable human capital, as well as demand for jobs, housing, education, income support, and social services.

Besides, both cities must try their best to cope with population aging as their populations are living much longer. In 2000, Shanghai's male and female residents could expect to live to 76.4 years and 80.5 years; for Hong Kong, the corresponding figures are 77.0 and 82.2 years. Growing longevity and low birth rates push up the proportion of the old population. In 2000, people who were 60 and older comprised 14.8 percent in Hong Kong.[8] In Shanghai, the aging trend arrived even earlier. Shanghai was the first Chinese city to attain the status of an aging society, defined as a society where 10 percent of the population was above 60 years of age or where 7 percent was aged 65 or above. In 1979, residents who were 60 and above reached

10 percent; in 2000, elders exceeded 18 percent (*Shanghaishi 2000 Nian Laonian Renkou Xinxi*, 2001; Gui, 2002). The challenge becomes graver due to such changing social and family structures as increased female labor participation, falling fertility, rising individualism, and shortage of urban housing.

On the economic front, Hong Kong and Shanghai have the largest economies among Chinese cities. In 2000, Hong Kong's GDP is HK$1,223 billion. Shanghai's economy is smaller: 455 billion yuan (HK$1 = 1.06 yuan) (*China Statistical Yearbook 2001*, pp. 57, 808, 821). It is important to note that both cities have to contend with deep structural transformations. In the case of Hong Kong, deindustrialization leaves behind rising numbers of older workers whose skills are not wanted by the new economy which is dominated by services, finance, and trade. Economic turmoil in Asia and Hong Kong's declining competitiveness worsened the problems of unemployment and poverty. Before 1997, unemployment was insignificant, at 2–3 percent. In the years following, the rate has risen sharply, to 6 percent in 1999, 4.4 percent in 2000, 6 percent in 2001, and 7.7 percent in April–June 2002 (*Ming Pao*, 17 July 2002, p. A2). Similarly, in Shanghai, market transition has thrown a lot of inefficient state enterprises and their employees into the gutters. Unemployment badly affects sunset industries such as textiles and machinery. Older workers, the unskilled, and the poorly educated become the casualties of the new mixed economy. Even though the official unemployment rate was only 3.5 percent in 2000, if redundant workers (*xiagang zhigong*) are also counted, the rate would be quite similar to Hong Kong's. Meanwhile, the weakening of the *danwei* system has tarnished the protective shield over urban employees and their dependents. As a result, many households fall into poverty and misery.

Changing economic and social circumstances call for new responses. Both cities are eager to better align the education system with the needs of the new economy. To that end, reforms in basic and higher education have been embraced with a vengeance. After successful consolidation of basic education, Hong Kong embarked on an ambitious expansion of higher education in the 1990s. In parallel there were reforms in the school curriculum, examination systems, and school management. In Shanghai, like other Chinese cities, the municipal authorities have been given more freedom to restructure courses, decide on student intake, improve on management, and generate funding in its schools and universities. Private educational institutions have likewise flourished to cater to the children of rich families. It would be interesting to compare and contrast the reforms in both cities and assess their effectiveness.

In addition to spending more to improve their human capital, both cities have invested heavily in town planning and housing in order to improve the physical infrastructure and ease overcrowding. The speeding up of housing development has been particularly notable. In Hong Kong, the government has successfully dispersed the population to the satellite towns. To promote home ownership, there has been increasing reliance on the building and sale of subsidized housing to the sandwich classes, as well as wider use of loans to help families buy flats in the private sector.

Table 1.1

Hong Kong and Shanghai, Basic Facts, 2000

	Hong Kong	Shanghai
Area:	1,098.04 square kilometers	6,340.5 square kilometers
Population:	Total population: 6,796,700 persons	Registered population (*huji renkou*): 13,216,300 persons
		Regular residents (*changzhu renkou*): 16,407,700 persons
Male:	3,309,700 (48.7%)	6,655,100 (50.4%) (*huji renkou*)
		8,430,300 (51.4%) (*changzhu renkou*)
Female:	3,487,000 (51.3%)	6,561,200 (49.6%) (*huji renkou*)
		7,977,400 (48.6%) (*changzhu renkou*)
Crude birth rate:	8.0/1,000	53/1,000 (*huji renkou*)
		55/1,000 (*changzhu renkou*)
Crude death rate:	4.9/1,000	72/1,000 (*huji renkou*)
		58/1,000 (*changzhu renkou*)
Natural increase:	20.9/1,000	−19/1,000 (*huji renkou*)
Population growth rate:	+1.5%	+6.5% (*huji renkou*)
Economy		
GDP:	U.S.$162.6 billion	U.S.$54.97 billion
Per capita GDP:	U.S.$24,392	U.S.$4,172.34
Labor		
Labor force:	3,406,000 persons	11,433,200 persons
Labor force participation rate:	Male: 1,966,000 (73.2%)	Male: 8,283,500 (78.27%)
	Female: 1,439,000 (49.2%)	Female: 7,452,400 (63.13%)
Unemployed persons:	151,000	200,800
Unemployment rate:	4.4%	3.5%
Underemployed persons:	91,000 persons	No such classification
Underemployment rate:	2.7%	No such classification
Education		
Kindergartens:	789	1,616
Students:	160,900	241,200
Primary schools:	816	1,021
Students:	494,900	7,886,000
Secondary schools:	525	1,120
Students:	466,700	1,052,800
Special schools:	74	34
Students:	9,400	5,400
HE institutes:	11	37
Students:	110,000	226,800

Table 1.1 *(continued)*

	Hong Kong	Shanghai
Health		
Hospital beds:	28,877	73,070
Bed ratio:	5.1/1,000	5.5/1,000
Doctor ratio:	1.5/1,000	3.8/1,000
Nurse ratio:	65.9/1,000	2.7/1,000
Social Security	Mandatory Provident Fund	Retirement Insurance (1999 coverage)
	Working employees: 1,664,000	Working employees: 4,374,000 Retired employees: 2,253,000
Social Assistance	Comprehensive Social Security Assistance: 228,000 cases	Minimum Security Allowance: 196,000 persons (*huji renkou*)
	Disability Allowance: 96,000 cases	No independent disability allowance
	Old Age Allowance: 452,000 cases	Retirement insurance (see above)
Housing		
Public rental housing:	676,000 (30.3%)	Comparable figures not available
Subsidized sale flats:	350,000 (15.7%)	Comparable figures not available
Private permanent housing:	1,205,000 (54.0%)	Comparable figures not available
Total housing units:	2,231,000	Comparable figures not available
Exchange rate	HK$7.791 = U.S.$1	8.280 yuan = U.S.$1

In Shanghai, new housing construction, redevelopment of the city center, promotion of market housing, and reform on the housing allocation system have completely altered the housing landscape. Not surprisingly, bold policies such as these have created winners and losers. They have also generated controversy and resistance, thereby posing daunting challenges for the state. The extent to which these policies fulfill their objectives would make interesting comparisons.

Rising demands amidst resource constraints calls for ingenuity in adjusting and improving the system of social protection. Both cities have to take a hard look at how they can better provide and fund the social services. There is need to reform their systems of health care, social security, personal social services, housing, and education. In so doing, the state in both cities has come to realize that onus for providing services should diversify. This means greater exploitation of market forces and civil society input in service production and management. Interestingly such approaches converge with global trends in curtailing the scope of the public sector and increasing the role of individuals in meeting their own needs. For China, the policy to pluralize responsibility has come to be known as the socialization policy. It has been suggested that this is akin to privatization strategies in the West (Wong, 1994, 1998). More recent research suggests that the exact forms this takes vary in different service sectors (Wong and Flynn, 2001). Across China on the whole, the socialization policy has become a dominant feature of the public goods regime. Social reforms of all sorts have increased both supply and quality. Likewise they have resulted in heavier burdens to consumers. It would be important to find out if this has been the case in Shanghai. Hong Kong has also had to grapple with similar issues in her own way. For instance, government has attempted to reform health care financing by tapping divergent sources of funding. This has been rigorously resisted so that until now Hong Kong's health care system still retains its "socialist" feature of universal provision at nominal charge. In another area, in order to halt sharp rises in social security expenditure, cuts have been made to welfare payments along with the appearance of work-fare schemes. However, further cuts in financial assistance, rent hikes in public rental housing, high tuition fees for associate degree courses, and increased charges for social services have proved unpopular. How Hong Kong handles these pressures will be instructive to Shanghai.

In our purview we have followed a broad conception of social policy. We believe a review of all key social programs is important to a holistic understanding of social policy in the two cities. Seven sectors comprising labor, education, health, housing, social security, the elderly, and immigrants are therefore included. Even then, it would be foolish to think we can deal with each policy area exhaustively. Hence we have allowed the authors to organize their discussion around themes and issues they considered most pertinent. These differential frameworks and perspectives should be obvious to the reader. Nevertheless, the theme of reform features prominently in all chapters. It is this commonality that frames the title of the book.

Over twenty researchers from Hong Kong and Shanghai have taken part in this comparative project. The Hong Kong team members are drawn from City University of Hong Kong while most of the Shanghai researchers come from East China Normal University. For each topic, Hong Kong and Shanghai researchers joined hand in hand. In terms of modus operandi, we asked the researchers to study pertinent development in their home city and jointly analyze the common and divergent features with their colleagues in the other city. The joint reflection and sharing came via joint field investigation in both cities and a series of workshops. The tangible outcome takes the form of two edited volumes. A book in Chinese detailing the development of social policy in the two cities has been published by East China Normal University Press in April 2003 (Gui and Wong, 2003).[9] Meanwhile, the current volume adopts a more eclectic approach. In each chapter, the lens is focused on selected themes, not a broad review of the entire policy area. There will be a necessary trade-off in breadth in order to sharpen depth, relevance, and poignancy. Toward a later stage of the project, Lynn White came on board by contributing to the editing and writing a commentary. With regard to funding, we are grateful to City University's Research Center for Comparative Public Management and Social Policy, later renamed the Governance in Asia Center. The Center's Flagship Project Grant allowed us to meet the expenses for fieldwork and discussion workshops.

In the rest of the chapter, a fuller discussion of the two cities' social policy development is presented. This allows the reader to appreciate the overall context, central features, and key tasks involved in the making of social policy during different historical periods. It also identifies the linkages across policies and explicates the interface between social policy and its environment.

Overview on Social Policy Development

Shanghai's Social Policy Chronicle

Research on social policy development in Shanghai is a fallow field. As far as we know, comprehensive studies of social development and social policy in the city do not exist. This applies to attempts at periodization. With little prior scholarship to draw on, we have to devise our own. Gui Shixun divides the evolution of social policy in Shanghai into five phases (Gui, 2003).

The first stage spanned the long period from 1983 to 1949. This phase began with the forced signing of the Treaty of Nanking in November 1843, which created the port of Shanghai. In 1845, the Shanghai Land Charter allowed the establishment of the British Concession. This opened the door to the creation of the French and American Concessions. These foreign enclaves on Chinese soil had enormous influence on jurisdiction and public administration. Their impact on political, economic, social, and cultural spheres was also profound. In 1927, the Nationalist (Guomindang) government gave Shanghai the status of a Special City and later promulgated a Big Shanghai Plan, marking the beginning of feeble attempts at city planning. The Japanese occupation of the foreign concessions from 1937 to 1945 and the brief spell of Guomindang rule did not succeed in restoring social order. The interregnum before communist victory was marked by chaos resulting from hyperinflation, mass poverty, and unemployment. All the while, Shanghai continued to attract migrants from its hinterland. When it first became a treaty port, the city had some 200,000 inhabitants. In 1949, the population had leapt to 5 million after successive waves of migrant influxes. Under Guomindang administration, welfare matters were divided up among the Bureaus of Internal Affairs, Public Welfare, and Social Affairs. In the concession territories, special agencies were set up to deal with health, relief, welfare, charity, and public cemeteries. These parallel systems operated side by side, each confined to its own area of jurisdiction. They could at best provide for emergencies and were completely inadequate in meeting needs. The more noteworthy attempts at managing social conflict were a series of labor regulations announced in 1927: Shanghai Labor and Capital Mediation Regulation, Shanghai Regulation on the Settlement of Industrial Conflict, and Temporary Regulation on Employee Pay Conditions. In the 1930s, the nationalist government went on to promulgate a host of labor laws, for example, factory law, trade union law, and labor-capital conflict resolution law. These laws prescribed the working age for children and apprentices, allowable work for women and child workers, minimum wage, working hours, freedom to form trade unions, holidays, subsidies, relief, maternity leave, prohibition on arbitrary dismissal and meddling with trade union affairs, and provision for industrial safety. The Social Affairs Bureau set up in 1928 was given authority over labor issues and industrial mediation. Because of territorial restriction and resistance of employers, the laws could not take effect in the foreign concessions. Even inside Chinese-held territories, enforcement was difficult when government authority and public finances remained weak and employers hostile (Zhang, 1990, pp. 825–40). Guomindang social concern for the welfare of workers was in harmony with the stance of the communists who conducted similar reforms in soviet-held areas.

The second stage began with communist victory and ended in 1956. The People's Liberation Army entered Shanghai on 27 May 1949. On the following day, the People's Government of the City of Shanghai was proclaimed. The new government immediately took over jurisdiction for defense, administration, finance, and education. In the economic sphere, an urgent task was to end the ruinous inflation that played a key role in toppling the Guomindang government. Enterprises owned by the Guomindang were taken into immediate state ownership. For a few years, the communists tolerated private capital. In 1956 the nationalization of industry, handicraft, trade, and commerce was completed. In the social domain, the state adopted policies to stabilize society, restore social order, and build a socialist culture. Hence the government lost no time in setting up work committees to relieve unemployed workers, repatriate vagrants, and administer work relief and worker training schemes. Immediate attention was also given to abolishing social evils such as prostitution, begging, narcotics, and crime. After a period of rehabilitation, the former "dregs of society" were assigned jobs or were resettled. Shanghai's burden in this regard was heavier than in other places because of its decadent past. The state busily set up reformatories for juvenile delinquents, criminals, and prostitutes. Reception homes were also run for the handicapped, orphans, and homeless elders. A particularly important gesture, in political terms, was an end to missionary-run schools and welfare programs, as local charities and welfare homes were operated by the Guomindang. By January 1956, eleven such agencies were turned over to the state. In addition, education work was completely revamped. Eleven higher education institutions were retained after completing reorganization, and 4 new institutes were established. To better meet the needs of new China, education cadres put emphasis on the teaching of science and technology, teacher training, and medicine. Public health work was enhanced as a matter of priority. In particular, campaigns to combat cholera, small pox, diphtheria, and typhoid were launched in a big way. Yet another key project was the introduction of retirement and health insurance systems for the urban proletariat under the Labor Insurance Regulations (1951, 1953) and Temporary Regulations on Retirement for State Employees (1955) promulgated by the State Administrative Council. Beneficiaries included employees working in state-owned enterprises, public-private joint enterprises, cooperatives, and government and party organizations. In 1956, relief schemes (the five guarantees) for childless elders and orphans living in the rural areas were introduced to overcome peasant resistance in joining rural collectives.

Gui called the third phase (1957–1965) the Stage of Exploration in Building Socialist Social Policy. In September 1956, Shanghai met the goals of the First Five-Year Plan a year ahead of schedule and thus completed the socialist economic transformation. In this period, the Soviet influence on Chinese development was substantial. Shanghai undertook economic construction on a scale and pace that was unprecedented. Nevertheless it was the Great Leap Forward that was to have an overriding effect in all spheres. In the social domain, Great Leap Forward policies plunged the city into chaos and hardship. Urban workers suffered reduced pay and saw an end to promotion, piece-rated pay, and bonuses. Rations for daily necessities such as food, cooking oil, and cloth were cut to reduce consumption. In 1961–1963, the city laid off 520,000 urban workers, sending 60 percent of them to villages outside Shanghai. The mass layoff was not enough to reduce the dependent population. Between 1963 and 1964, 41,000 educated youths were mobilized to join the Construction Corps in Xinjiang. The expressed purposes were to lesson the pressure on urban employment and food supply and reduce state expenditure on wages (Zhonggong Shanghai Shiwei Dangshi Yanjiushi, 1999, pp. 200–228).

The biggest mistake in social engineering lay in the urban people's commune movement. In August 1958, the Political Bureau of the CCP passed the resolution to set up people's communes in villages. Rural people's communes were given the remit for industry, agriculture, commerce, education, and defense matters to function as the basic tier of administration. Chairman Mao's dream was that the policy would allow China to attain communism in three to four years, or five to six years at the most. A month later, Shanghai obeyed the chairman's call. The government set up the first rural people's commune, the 7–1 People's Commune, by amalgamating seven townships and villages. By October, the communization of Shanghai's suburban areas was concluded with the formation of twenty-three communes. These organizations nationalized all political, economic, administrative, and distribution functions. Some even provided free canteens, health care, schools, and relief for the needy. For example, in mid-1958, Shanghai's rural communes set up 186 old people's homes catering to 6,461 residents (*Zhongguo Minzheng Tongjian Nianjian 1993*). The frenzy to leapfrog into communism was extended to the city areas. The relevant trials began in April 1960. The dependents of urban workers and laborers outside the state and collective sectors were mobilized to join urban people's communes. The main activity of these entities was organizing industrial enterprises and life-support services such as canteens, nurseries, and literacy classes. As is well known, this movement was a fiasco. By 1961, the government was forced to carry out a program of "adjustment, consolidation, filling out and improvement" (Cheng, 1986, p. 30). In the countryside, there was a return to household cultivation, an end to arbitrary egalitarian distribution, and greater devolution in decision making. Most of the old people's homes that sprang up suddenly were shut down almost over night. By the end of 1962, only ninety rural welfare homes catered to 957 residents. Likewise, the urban people's commune movement came to an end in September 1962 (Zhonggong Shanghai Shiwei Dangshi Yanjiushi, 1999, pp. 200–228).

The fourth phase coincided with the Cultural Revolution (1966–1976). Gui dubbed this the Period of Radical Left Social Policy. Because Shanghai was the main base for much of the machination of the Gang of Four, the city became a key disaster area of the fanatical movement. During the decade, its economy was gravely disrupted. Governance descended into anarchy. Living standards went down sharply. In these years, the city lost at least 20 billion yuan in revenue. Average monthly wages in state-owned enterprises declined from 70 yuan in 1965 to 57 yuan, a fall of 18 percent (Zhonggong Shanghai Shiwei Dangshi Yanjiushi, 1999, pp. 365–67). The city's schools and universities could not escape severe setbacks. With professors and intellectuals denounced as members of the "stinking ninth-category," classes were suspended. Of the city's twenty-four higher education institutions, only sixteen remained after a spate of closure, merger, and relocations. College intake was halted for four years so that in 1971 it shrank to 3,510 freshmen, from some 52,000 in 1965. Annual admission resumed in the 1970s. However, university entrance examinations were scrapped in favor of admitting workers, peasants, and soldiers, wrecking the quality of intake and normal teaching. Middle-level vocational education met a worse fate. All fifty-one vocational middle schools stopped recruitment. Technical skill training and agricultural middle schools were closed, putting a virtual stop to professional and technical education. An even bigger disaster for young people was the rustification program. Under its banner, a total of 1,017,000 high school students from the city were sent down to the countryside. This put an abrupt halt to their education and marred their long-term prospects. Urban planning and management suffered as well. For five years, not a square meter of new housing was constructed. In the entire decade, only 1 million square meters of living space was being built. This caused severe housing shortage and overcrowding when the city emerged from the lost decade. For example, in 1975, a third of all households in the city had a per capita living space below 4 square meters. Many married couples could not get housing and had to cram in with their parents. In social welfare matters, the only achievements were some facilities for the mentally ill and work schemes for the handicapped. By and large, the bulk of welfare work languished and stagnated.

The end of the Cultural Revolution marked the dawn of a new age. Gui called this the Era of Comprehensive Social Reform. The arrest of the Gang of Four and the Third Plenum of the 11th CCP Central Committee (December 1978) saw China turn a new chapter in her development saga. The leaders announced the end to class struggle and adoption of an open door and reform policy. As is well known, economic reforms began in the countryside and then spread to the urban areas. China made every effort to attract foreign capital, technology, and management know-how and to expand trade and export. The adoption of open door policies culminated in China's entry into the World Trade Organization in 2001. In the space of twenty years, the Chinese economic system has transformed itself into a market economy, albeit with socialist remnants. The entry into the World Trade Organization in 2001 concluded China's vision to integrate fully with the global economy. The paradigm shift from plan to market requires reengineering beyond the reach of the economy. Reforms in the spheres of politics, administration, science and technology, education, culture, and social security were taken on board. The outcome was a complete shake-up of China's social systems.

Shanghai's attempts at social restructuring were aimed at rectifying the havoc and destruction inflicted by the Cultural Revolution. Education reform got early attention. In 1977, the government revived entrance examinations for university admission. In 1985, the management of schools was divided among different tiers of administration. From 1992 onward, state monopoly in education was relaxed to allow people-run schools and joint public and society–run schools to operate. In 1993, universities and colleges in Shanghai were classified into centrally administered and locally administered establishments. Besides, university graduates were no longer guaranteed a job or promised a cadre post, which previously had been considered their right. Like everybody else, they have to try their luck in the open market.

Shanghai also led the country in piloting reforms in social security. In 1984, the city introduced pooled collection and distribution of pensions, paving the way to change enterprise-run retirement schemes into socially administered pensions. In 1994, the Retirement Insurance Regulations for Employees in Cities and Towns in Shanghai were announced. This was followed by Regulations on Society-Run Retirement Insurance for Rural Elders in Shanghai in 1996. These regulations heralded a universal retirement scheme for all urban employees in 1997. Similarly, health insurance reforms started in 1992 and in the following years continued to review and improve coverage for inpatient and outpatient care for urban employees and to restore cooperative health insurance in rural areas. In 2000, the Regulations on Basic Health Insurance for Workers in Cities and Towns in Shanghai were announced to offer coverage of basic health care for all urban employees. As far as unemployment insurance is concerned, reform efforts started in 1986 and continued right through to 1995 with the announcement of the Regulations on Unemployment Insurance in Shanghai. To facilitate reemployment of displaced workers, Shanghai set up Re-employment Service Centers in 1996 (Chen, 2000; Lee, 2000). These function as halfway reception centers to simultaneously provide retraining, job introduction, living allowance, and social security for redundant workers before they are ready to find jobs in the open market. These centers are considered so useful that the central government promotes them across the country. Shanghai's attempts to facilitate informal sector employment and job schemes for middle-aged workers (the 40–50 Employment Project) are also highly praised. These employment projects attract a steady stream of visitors who come to learn from Shanghai's example.

In the field of housing, Shanghai set out to expand housing construction to make up for missed opportunities. Between 1979 and 1985, 25.5 million square meters of housing space was built, equivalent to the total stock constructed in the first twenty-nine years (Zhonggong Shanghai Shiwei Dangshi Yanjiushi, 1999, p. 103). At the same time, the inner-city areas were cleared for redevelopment, often in collaboration with private developers. Displaced residents were resettled to outlying areas in more spacious but less conveniently located housing, spawning strong complaints. To help local residents buy their homes, the city set up a housing provident fund.

As far as social relief reform is concerned, Shanghai has been an acknowledged leader. In 1990, the city introduced subsidies to help destitute households obtain emergency medical treatment. As the problem of urban poverty worsened due to rising unemployment and enterprise difficulties, the city responded by revamping its system of financial assistance. In 1995, Shanghai became the first Chinese city to give out food and cooking oil cards to destitute households. A year later, the government invented a subsistence protection allowance scheme, *zuidi shenghuo baozhang*, for residents living below the local poverty line (Tang, 2001). This scheme was later replicated across the country. Other worthwhile attempts to help vulnerable groups included the promotion of community services run by neighborhood organizations and services for the elderly. To cope with the rising numbers of temporary residents, the government also passed a number of regulations to register, manage, and set restrictions to the type of jobs that could be taken by out-of-towners.

Hong Kong's Social Policy Chronicle

A number of Hong Kong scholars have written on the history of social policy in Hong Kong. They include Peter Hodge (1976, 1981), Nelson Chow (1980, 1984), and Raymond Chan (1996). Another useful contribution comes from Tang Kwong-leung whose periodization we will follow here. Basically Tang divides the evolution into five stages (Tang, 1998). The first phase covers the period between 1842 and 1952. Tang calls this the Period of Residual Social Policy, residual in the sense of being of secondary importance, vis-à-vis for example, economic development. In this long gestation period, social policy in Hong Kong was marked by an absence of clearly conceived and comprehensive agenda of social development. To Britain, the business of Hong Kong was business. The major, if not sole, purpose of running a colony was to exploit maximum economic benefits for the metropolitan country, not to create a welfare community. In the early days, Hong Kong's colonial masters were particularly opposed to improving welfare provision out of fear that huge influxes of poor people from south China would swamp the territory, thereby imposing an unbearable burden on Hong Kong and Britain. Another concern stemmed from the permeability of the Hong Kong-China border. Before the Second World War, people could move freely between the two territories. The two-way human flows underscored the following argument: since Hong Kong did not have a stable population, there was no way to determine who were the true citizens with social rights. There was the additional excuse that improving welfare for the local population would undermine Chinese culture. The fine tradition of self-reliance, family responsibility, and local philanthropy aid would thus weaken if Western models of welfare were transplanted to Hong Kong. Such a defensive position has been severely criticized by scholars and social critics (Hodge, 1981; Chow, 1980; Chan, 1996; Lee, Chiu, Leung, and Chan, 1999). Politically, what may be more relevant is the lack of popular demands. Many Hong Kong residents regarded Hong Kong as a place of refuge. New arrivals from China had never enjoyed any entitlement to state support and had low expectations of what to expect from the government. Their mind-set of endurance and quiescence led them to search for private salvation rather than making demands on the state (Hodge, 1981). Destitute people without family support turned to neighborhood associations, local charities like the Tung Wah Group of Hospitals and Po Leung Kuk, and overseas missionary bodies for emergency relief (Sinn, 1989).

The first century of colonial rule came to an abrupt end when Japan occupied Hong Kong. In this period of three-years-and-eight-months, many local residents escaped to their native villages in Guangdong. After the war, masses of returnees poured back. Between 1945 and 1946, the population jumped from 600,000 to 1.6 million. The success of the communist revolution brought a further avalanche of Mainland migrants. The population explosion caused gigantic problems. The immediate aftermath was a scarcity of goods and provisions of all kinds, including food. Jobs and housing shortages were also critical. The images of poor people crammed into decrepit tenement blocks, whole families to one room, slumming on hillside huts, and bedding down on the streets, rooftops, and staircases were powerful reminders of appalling living conditions at the time. Not surprisingly, severe overcrowding led to outbreaks of infectious diseases, particularly tuberculosis, and public hazards such as poor sanitation, crime, and fire. It was the last scourge that changed the government's attitude to social intervention. On Christmas Eve 1953, a massive fire broke out in the squatter huts of Shek Kip Mei, destroying the "homes" of over 50,000 people. Government response to the crisis marked the turning point of laissez-faire policy.

Tang christened the second phase the Period of Partial Social Policy. This extended from 1953 to 1970. During this time, the attitude of the government underwent a radical shift. The squatter fire at Shek Kip Mei was the tip of the iceberg where the housing issue was concerned. To resettle the victims, the authorities set out to build emergency shelters. This was followed by a review of housing needs and a massive public housing program to build low-cost rental dwellings for poor families. The continuous inflow of Mainland migrants multiplied demands for social services such as education, health, and social welfare. Widespread poverty, harsh living conditions, and service shortages fanned social tensions. Political conflicts occurring in China also produced ripples of instability in Hong Kong society.

The 1950s and 1960s were a period of incipient labor unrest. Wages and working conditions were poor. Worker frustrations had been bottled up because of high competition for jobs and the indifference of employers. At critical junctures, however, pent-up grievances erupted into violence under the instigation of external political influences. The labor movement in this period was the scene of intensive mobilization by unions belonging to pro-China and pro-Taiwan political camps. The riots that broke out in 1956 and in 1966–1967 could be traced to forces that sought to exploit dormant labor conflicts and to link them with political agitations outside the territory. Public dissatisfaction with a government seen as out of touch with the hardship of the common people eased ignition. The 1966–1967 riots, in particular, have had a profound impact on Hong Kong society and governance. Riding on the wave of Cultural Revolution fanaticism, local leftists sought to mobilize Hong Kong workers to support the struggle in China, strike out against the capitalists and government, and bring an early return of the colony to the Motherland. The unprecedented chaos that broke out frightened government and society into united action in order to restore order and social solidarity. Thereafter government adopted many measures to improve communication with the people and strengthen public consultation. Activities such as the Hong Kong Festival, recreational programs for youths, and the Clean Hong Kong Campaign were mounted almost immediately. More important, the government began in earnest to examine the community's expectations and to foster a sense of local identity. Social services planning was stepped up to meet the needs of the people in a more affluent society (Chow, 1980; Chan, 1996). Social stability and social integration became primary concerns that translated into active social policy programs in the next stage.

The Stage of Big Bang Social Policy followed next, spanning the period between 1970 and 1978. In 1972, Hong Kong had a new governor. Sir Murray MacLehose's arrival signaled an end to the noninterventionist stance of the state in social affairs. Instead, the government became an active service provider. In 1973, the first Five-Year Plan for Social Welfare Development was published. The same year also saw the announcement of an ambitious housing plan with the aim to build homes for 1.8 million families in ten years. In a parallel development, a massive new town-building program began, to disperse the population from the city center and to improve the living environment. As far as education was concerned, there was much catching up work to do as befitting an industrial society. In 1979, nine years compulsory schooling for children between six and fifteen was introduced, on the basis of six years free primary education achieved in 1971. The magnitude and pace of these initiatives were so stunning that many commentators called the MacLehose years The Golden Age of Social Policy.

The social welfare five-year plan marked an important milestone for welfare planning. It was the first time the government acknowledged it had ultimate responsibility for welfare and spelled out its objectives and future directions. The contribution of voluntary welfare agencies was recognized, giving them partnership status alongside government. In about a decade (1966–1977), social service allocations (education, health, housing, and social welfare) increased sixfold, from $600 million to $3.73 billion. This translated into an average growth of 30–40 percent per year and 5–6 percent of GDP during the period. On the social welfare program alone, the budget surge was even more striking. In 1970, welfare services outlay was a modest $37 million; in 1977, this jumped to $393 million, a tenfold increase. In terms of GDP, the share jumped from 1.51 percent to 4.37 percent (Tang, 1998, pp. 72–74). A similar pattern was seen in the education area. In 1972, government expenditure was $845 million. By 1978, this had gone up to $1.981 billion. In health care, too, the government embarked on a major expansion of hospital beds, specialist services, and manpower. By 1978, health spending rose to 8 percent of public expenditure (Tang, 1998, pp. 61–67). All in all, big advances in the social programs went a long way in healing the social wound in an underprovided society. More important, government began to see the deeper meaning of social programs beyond their mundane purposes. Conceptually, Hong Kong was likened to a building, and education, health, housing, and welfare were seen as the four pillars of society. Investing in these sectors was crucial to reducing social conflict, increasing political legitimacy, and enhancing social integration, thereby contributing to community-building. When the 1970s ended, Hong Kong emerged as a society with a fairly elaborate welfare state. In retrospect, these gains could not have come about without rapid growth of the economy. Still, one would argue that the will to reform derived largely from concern for regime stability. Hong Kong has certainly learned its lesson from the mid-1960s riots. When Sino-British negotiations on Hong Kong's future commenced at the end of the decade, Hong Kong citizens developed a strong sense of local identity.

Tang called the next stage the Period of Incremental Social Policy. This phase began in 1978 and ended in 1997. The dominating ethos of the time was the Sino-British negotiations. Public and government preoccupation with transition concerns eclipsed everything else. In social policy matters, an incremental agenda carried the day. This meant that apart from small-scale improvements in existing programs and modest expansion to meet growing demands, policy continuity rather than bold changes was the norm. After completing the social project of the MacLehose years, big leaps in innovation and spending was judged unnecessary. Still, social spending grew at a brisk rate of 10 to 20 percent per year during this period although the momentum had slowed when compared with the previous decade. Between 1978 and 1992, public expenditure on the social services increased from 3.68 percent to

Table 1.2

Social Service Spending in Hong Kong as Proportion of Public Expenditure (%)

	1994/95	1995/96	1996/97	1997/98	1998/99	1999/00
Education	17.4	17.6	17.9	20.0	18.5	19.0
Housing	11.9	10.0	11.5	10.5	14.8	16.0
Health	11.6	12.7	11.9	11.9	11.3	11.3
Social welfare	6.6	7.4	8.5	9.3	9.8	10.6
Total	47.5	47.7	49.8	51.7	54.4	56.9

Source: Hong Kong Government, *1999/2000 Government Budget*, Hong Kong: Government Printers, 1999.

6.29 percent. The sector that stood out in terms of spending boost was social welfare, rising from 6.6 percent of public expenditure to 9.8 percent. The pattern of development can be gleaned from Table 1.2.

Incrementalism in social policy was indicative of a more cautious approach to public finance after the big bang push of the 1970s. All along, the government has maintained abiding commitment to fiscal prudence. One may even argue that the MacLehose exuberance was an exception rather than the norm. The central plank of government philosophy has always been to link public expenditure with economic growth. After the change of sovereignty, the government has been even more anxious to return to conservatism and fiscal sustainability. The preservation of Hong Kong's low taxation regime, seen as the cornerstone of its economic success, is represented as a core prerogative. The Chinese side is particularly keen to safeguard the integrity of Hong Kong's reserves and avoid external debt. This position was later enshrined in the Basic Law of the future Hong Kong Special Administrative Region (Ghai, 1997; Wesley-Smith and Chen, 1988). Frankly China did not want Britain to steer Hong Kong into a welfare state. The sharp rise in social security spending was already a shock to Beijing. The danger of Western-style profligacy may be exaggerated. Hong Kong's big push for social development in the last decade was built on a low base. By 1995, public expenditure accounted for only 17.8 percent of GDP. This was decidedly modest compared with 33.6 percent in the Netherlands, 31.0 percent in Germany, and 27.3 percent in Britain (Commission of the European Communities, 1995, p. 60; Flynn, 1997, p. 25; Hong Kong Information Services Department, 1999, p. 437). There is a long way to go before Hong Kong can "catch up" with industrial societies at similar levels of prosperity.

The final stage covered the period after the 1997 reunification with China. Even before the transition, the social and economic situation had undergone very considerable changes. The open door and reform program in China galvanized an exodus of Hong Kong capital and industry to take advantage of cheap wages, low land and production cost, and huge domestic market over the border. This process started in the 1980s. By the mid-1990s, deindustrialization was virtually complete. The transformed economy came to be dominated by finance, real estate, commerce, and professional and tertiary services. The Asian financial crisis in 1997 dealt Hong Kong a severe blow. Immediately following the change of sovereignty, the economic shock greatly increased social instability. The severity and prolonged nature of the onslaught caught the government off guard. Unlike her trading partners in the region, Hong Kong resisted devaluation. The strength of the Hong Kong dollar compounded the problem of competitiveness. Despite the fact that wages and land prices had fallen drastically since 1997, the end to recession was nowhere in sight. In corollary, unemployment soared to new heights, from 2–3 percent before 1997 to over 6 percent in 2000. Needless to say, workers became the main victims. The poor state of the economy had another effect—it triggered a fall in general revenue. Receipts from the sale of land, which previously accounted for a quarter to a third of state income, fell sharply. Sustained structural deficits stirred the fear that the reserves could be wiped out in five to six years if the economy did not rebound and public spending remained at high levels. The government and the business-dominated legislature were deeply worried.

Without any sign of an economic recovery, public opinion demanded fiscal prudence. The government was forced to look for ways to save money and raise revenue. Reforms of the public sector and the civil service advanced to the top of the public agenda. Reviews on social security to deter welfare dependency came largely from concerns over the rise of social assistance spending on the unemployed and single-parent families. Health financing reform can be interpreted in this light as well. With the elderly population growing quickly, doubts were also raised about the sustainability of universal health care at nominal charges. The government turned to review health care financing as a response to the dilemma.

The pressure for reform did not come solely from economic exigencies. A number of policy blunders occurring shortly after the handover stirred strong public anger. The incidents included the chaos during the opening of the new airport, poor management of the chicken flu epidemic, and the scandals linked to faulty piling work at a housing estate in Tin Shui Wai. What these cases revealed were poor coordination, inflexibility, and failure in the bureaucratic and monitoring systems. Perhaps the biggest fiasco was the chief executive's housing plan. Determined to bring down the pre-1997 house prices and speed up housing construction, Tung Chee-hwa trumpeted a bold plan to build 85,000 housing units per year and promote home ownership for 70 percent of families. The so-called "85,000 policy" was roundly condemned for the catastrophic collapse of property prices. Finally the plan was quietly abandoned. To shore up the housing markets, the government temporarily suspended land sales and curtailed the supply of public housing. In education, the reforms are designed to address qualitative issues, such as improving management of schools, making better use of resources, and promoting information technology. The purpose of these reforms is to align education more closely to the needs of the new knowledge society. To help Hong Kong compete in an increasingly globalized economy, improvements in education have been given urgency and priority.

All in all, civil service is facing unprecedented challenges. Before the handover, it was praised as one of the best in the world. Now the public finds it wanting in many aspects and demands better performance, greater accountability, and reduced cost. The chief executive and his advisors are also under pressure to talk less and do more. As unemployment worsens, the public wants more resolute measures to create jobs. The community is also united in demanding more effective leadership to lead Hong Kong out of recession and identify its future direction. Meanwhile, social tensions keep surging. People demand help to manage the risk of social and economic transition. The challenge to maintain and improve Hong Kong's social protection systems against economic and political constraints will continue to frame the agenda for public debate and social policy in the years to come.

A Brief Summary of Themes

All the chapters that follow address the common theme of policy reform. Each, however, focuses on specific issues. Chapter 2, by Anthony Cheung and Gu Xing-yuan, has its eye on health care finance reform in the two cities. Both health care systems were dominated by a strong state role. Since the 1980s, both systems have actively contemplated or embraced reform, in particular in the mode of financing to cope with severe resource constraints. Their respective change processes reveal dilemmas among the objectives of financial viability, equitable coverage, and citizens' affordability in paying for health care.

In the next chapter, Kwok-yu Lau, James Lee, and Zhang Yongyue identify the enormous changes in the housing systems in Shanghai and Hong Kong. Similarities and differences in policy changes are apparent. In particular, signs of convergence in tenure structure are becoming more evident. The review concludes that an appropriate institutional framework and effective regulatory systems will have enormous importance in the formulation of successful and sustainable housing policy.

In the chapter on education, David Chan, Ka-ho Mok, and Tang Anguo discuss the reforms in education that were carried out in the two cities. In particular, the strategies adopted in the 1990s and key policy indicators are examined in detail. The effect of globalization, decentralization, and marketization in the education systems is discussed and compared.

The chapter by Grace Lee, Wang Daben, and Simon Li begins with an overview of the labor market conditions and trends in employment in Shanghai and Hong Kong. It then analyses the causes and extent of unemployment in both cities. State responses to the unemployment problem in the form of employment policies and active labor market programs are evaluated.

Poverty and social security reform are the themes of the chapter by Raymond Ngan, Ngai-ming Yip, and Wu Duo. The discussion begins with an overview of the political-economic context that necessitates a revamp of social security. The focus is placed on social security programs to help the unemployed poor. It ends with an assessment of program effectiveness in the two cities.

The contribution from Alice Chong, Alex Kwan, and Gui Shixun takes a broad look at the patterns of population aging, as well as welfare provisions for frail elders in the two cities. The achievements and pitfalls of relevant programs are compared. The authors believe that the common challenges facing both have to do with increasing demand for care, weakening of the caring ability of families, and shortages in service provision. To rectify the shortcomings, eleven policy recommendations have been proposed.

Ray Yep, King-lun Ngok, and Zhu Baoshu's chapter focuses on the theme of state, market, and immigrants. Despite their characteristic as immigrant communities, local state policies in regulating immigration and utilizing the contribution of immigrants differ. The authors suggest that the Shanghai approach is more flexible and effective than Hong Kong's.

In the final chapter, Lynn White offers a commentary on the two cities' approach to urban planning, business development, and policy making and their differences. The three major tensions—accumulating capital or financing welfare, metropolitan adjustment to inland development, and capitalization with compassion—that both face are found to have special salience for social policy and development.

Notes

1. Also, according to Yao Xitang, Shanghai had remitted 412 billion yuan out of a total gross national income of 477 billion yuan in the first forty years under communism. See Yao Xitang, ed., *Shanghai Xianggang Bijiao Yanjiu* (Shanghai and Hong Kong Comparative Study) (Shanghai: Shanghai Renmin Chubanshe, 1990), p. 2.
2. There are many good general works on Shanghai. Some of these include Y.M. Yeung and Sung Yun-wing, eds., *Shanghai: Transformation and Modernization under China's Open Policy* (Hong Kong: Chinese University Press, 1996); Rhodes Murphy, *Shanghai: Key to Modern China* (Cambridge: Harvard University Press, 1953); Christopher Howe, ed., *Shanghai: Revolution and Development in an Asian Metropolis* (Cambridge: Cambridge University Press, 1981); Betty Peh-T'i Wei, *Shanghai: Crucible of Modern China* (Hong Kong: Oxford University Press, 1987); Lynn T. White, *Shanghai Shanghaied? Uneven Taxes in Reform China* (Hong Kong: Center of Asian Studies, University of Hong Kong, 1991); Chen Yi, ed., *Dangdai Zhongguo de Shanghai* (Shanghai in Contemporary China), Vols. I and II (Beijing: Dangdai Zhongguo Chubanshe, 1993); Zhongguo Gongchandang Shanghai Shiwei Dangshi Yanjiushi, ed., *Zhongguo Gongchandang Zai Zhongguo 1921–1991* (The Chinese Communist Party in Shanghai 1921–1991) (Shanghai: Shanghai Renmin Chubanshe, 1991); Zhonggong Shanghai Shiwei Dangshi Yanjiushi, ed., *Shanghai Shehui Zhuyi Jianshe Wushinian* (Fifty Years of Socialist Construction in Shanghai) (Shanghai: Shanghai Renmin Chubanshe, 1999); Shanghaishi Jingji Xuehui, ed., *Kaifa, Kaifo, Kaifang—Shanghai Jingji Fazhan Zhonglun* (Development, Port Founding and Opening—Comprehensive Commentary on the Economic Development in

Shanghai) (Beijing: Zhongguo Guangbo Dianshi Chubanshe, 1992); Marie-Claire Bergere, *The Golden Age of the Chinese Bourgeoisie 1911–1937* (Cambridge: Cambridge University Press, 1989); Zhang Zhongli, *Jindai Shanghai Chengshi Yanjiu* (Urban Studies on Modern Shanghai) (Shanghai: Shanghai Renmin Chubanshe, 1990).

3. Good general works on Hong Kong include the following: G.B. Endacott, *Government and People in Hong Kong* (Hong Kong: Hong Kong University Press, 1964); Peter Wesley-Smith, ed., *Hong Kong's Transition: Problems and Prospects* (Hong Kong: Faculty of Law, University of Hong Kong, 1993); Siu-kai Lau, *Society and Politics in Hong Kong* (Hong Kong: Chinese University Press, 1982); Norman Miners, *The Government and Politics of Hong Kong* (Hong Kong: Oxford University Press, 1991); Ian Scott, *Political Change and the Crisis of Legitimacy* (Hong Kong: Oxford University Press, 1989); David Faure, *Society: A Documentary History of Hong Kong* (Hong Kong: Hong Kong University Press, 1997); David Bonavia, *Hong Kong 1997: The Final Settlement* (Hong Kong: SCMP, 1985); Wong Siu-lun and T. Maruya, eds., *Hong Kong Economy and Society: The New Era* (Hong Kong: Center of Asian Studies, University of Hong Kong, 1998); and Jan Morris, *Hong Kong: Epilogue to Empire* (London and New York: Penguin Books, 1997).

4. In 1995, per capita GDP in Hong Kong was U.S.$22,618. See *China Statistical Yearbook 2001*, p. 808.

5. A good recent work is Yang Tuan's *Shehui Fuli Shehuihua: Shanghai yu Xianggang Shehui Fuli Tixi Bijiao* (Social Welfare Socialization: Comparative Study on the Social Welfare Systems of Shanghai and Hong Kong), edited by Yang Tuan and published by Huaxia Chubanshe, Beijing, 2001. The focus is on social welfare only.

6. Source: http://news.eastday.com/epublish/big5/paper148/20010403/class014800006/hwz 352923.htm.

7. According to Nelson Chow, between 1977 and 1980 some 300,000 to 400,000 Mainland immigrants came to Hong Kong. See Nelson W.S. Chow, *Shehui Baozhang he Fuli Zengyi* (Controversies in Social Security and Welfare) (Hong Kong: Hong Kong Tiandi Tushu Youxian Gongsi, 1994). Before 1980, an illegal immigrant who succeeded in entering the urban area of Hong Kong could register for Hong Kong residency under the so-called Touch Base Policy. After 1980, this policy was replaced by the Instant Arrest and Instant Repatriation Policy. Mainland immigrants could apply for family reunion in Hong Kong under a daily quota of seventy-five persons. This quota was later increased to 105 and 150. Adding up these figures means that some 800,000 to 900,000 Mainland immigrants have come to settle in Hong Kong between 1977 and 1996.

8. Source: www.info.gov.hk/censtatd/eng/hkstat/fas/pop/by_age_sex_index.html.

9. The publication is enentitled *Shanghai yu Xianggang Shehui Zhengce Bijiao Yanjiu* (Comparative Study on Social Policy in Shanghai and Hong Kong), edited by Gui Shixun and Linda Wong (Shanghai: Huadong Shifan Daxue Chubanshe, April 2003).

—— 2 ——

Health Finance

Anthony Cheung and Gu Xingyuan

Introduction

The mainland of China and Hong Kong have historically pursued different economic and social systems, one rooted in socialism and the other in capitalism. In recent years, China has moved toward a so-called socialist market economy, with social services such as housing, education, and health care increasingly marketized, privatized, and "societized" (*shehuihua*, i.e., the process of returning the responsibility to society through community financing rather than state financing). This trend may render some degree of convergence between the mainland and Hong Kong systems. In health services, however, the two systems have shared some similarity in that while state organs, enterprises, and urban work units in the mainland have all along been part of a state-funded health care system,[1] Hong Kong's public health care system also has strong features of a "national health service"— with the government providing extensive public health care to the population. Since the 1980s, both systems have undergone reforms and changes.

Current health care finance reforms in both Hong Kong and Shanghai have been triggered by financial constraints. The respective reform processes illustrate conflicts among objectives of financial viability, equitable coverage, and citizens' affordability to pay for their health care. This chapter discusses the trajectories of reform in these two major cities of China against the background of their diverse legacies; it concludes with a comparative overview.

Hong Kong: Problems of Financial Sustainability

Institutional Setting

Health care services in Hong Kong were previously run by the Medical and Health Department (MHD) which had grown from a very humble beginning as a medical service for soldiers and colonial civil servants in the mid-nineteenth century into

Table 2.1

Hong Kong's "Dual System" of Health Care

	Public sector (HA, Department of Health and other institutions)	Private sector (private hospitals and GPs)
Primary care	30%	70%
Secondary and tertiary care	92.1%	7.9%
Extended and long-term care	100%	Nil
Distribution of hospital beds in 1999	91.1%	8.9%

Source: Hospital Authority (2001, 2002). Primary care refers to outpatient, community health, and preventive services. Secondary and tertiary care refers to hospital-based medical treatment and rehabilitation services.

one of Hong Kong's superdepartments by the 1980s. The importance of MHD grew with the adoption by the government of the first Ten-Year Medical Development Plan in 1963, followed by the 1974 White Paper on *The Further Development of Medical and Health Services*, which paved the way for massive expansion in the provision of low-cost or free health care services to the population. In 1985, when the government commissioned the consultant W.D. Scott to review the delivery of public hospital services, MHD had a staff of 24,330 out of the total civil service establishment of 181,720 (or 13 percent) and controlled over 9 percent of the government budget (Hong Kong Government, 1985). It directly managed 55 percent of the 21,337 hospital beds in the public sector. The rest were hospitals operated by nongovernment sponsoring bodies under government subvention and MHD's supervision (known as "subvented" hospitals). After the setting up of the Hospital Authority (HA) in 1990, it took over all government and subvented hospitals, with primary and preventive care provided by a new Department of Health.

Historically Hong Kong had developed some kind of "dual system" whereby public hospitals made up the bulk of hospital care, while the private sector provided most of the outpatient consultation. Table 2.1 gives the latest picture.

The public sector is tax funded, and users are only required to pay a nominal fee. Treatment at a government outpatient clinic costs HK$37 a visit, which includes medicine, as well as X-ray examinations, laboratory tests, and so on. Specialist outpatient consultation costs HK$44. At public hospitals, patients are only charged HK$68 per day for a stay in the general wards when hospitalization is required; this amount covers diet, medicine, X-ray, and various tests. For those patients on public assistance, all charges are waived. The government has long

Table 2.2

Basic Health Indicators of Hong Kong Compared with Selected Countries, 1996

	Life expec- tancy (male)	Life expec- tancy (female)	Infant mortality (per 1,000)	Maternal mortality (per 1,000)	Health expendi- ture (% GDP)	Public health expendi- ture (% GDP)	Private health expendi- ture (% GDP)
Hong Kong	**76.0**	**82.0**	**4.0**	**7.0**	**4.6**	**2.5**	**2.1**
Singapore	75.0#	79.2#	3.6#	0#	3.2*	1.2#	2.3*
Japan	77.0	83.3	3.8	n.a.	7.2	5.7	1.5
Australia	75.2	81.1	5.8	9.0*	8.5	5.9	2.6
U.S.	73.0	79.0	7.2	7.5	14.0	6.5	7.5
U.K.	74.4	79.3	5.9	9.0*	6.9	5.8	1.1

Source: Harvard Team (1999, p. 49, Table 4.16).

* = 1995 figures; # = 1997 figures

maintained a policy of not preventing citizens from receiving appropriate care and treatment because of a lack of economic means.

Achievements and Problems

The merits of Hong Kong's system are

- It is relatively cost-effective and budget-capped, helping to contain overall medical and health expenditures.
- It is equitable, with access to health care that is available to all of the population, rich or poor, not unlike a socialist system.
- Because of its nominally "free" service, the ability to pay is basically not a problem on the part of patients.

The efficacy of the system has been borne out by health outcomes. Life expectancy at birth was 77.2 years for men and 82.4 years for women in 1999 (Hong Kong Special Administrative Region Government, 2000). The infant mortality rate was 3.2 per 1,000 live births in 1999 (Hong Kong Special Administrative Region Government, 2000), down from 19.0 in 1970, while the maternal mortality ratio per 100,000 live births was 7.0 over the 1990–1996 period (Harvard Team, 1999, p. 23). Hong Kong's basic health indicators compared favorably with other developed countries' despite Hong Kong's low health care expenditure in terms of GDP (see Table 2.2).

In December 2001, the total number of hospital beds was 35,860, and the total number of registered doctors was 10,412. The bed-population ratio and doctor-

Table 2.3

Hospital-Bed-Population and Doctor-Population Ratios of Hong Kong Compared with Selected Countries

	Beds per 1,000 population	Doctors per 1,000 population
Hong Kong	**5.1 (2001)**	**1.5 (2001)**
Japan	13.1 (1998)	2.0 (1998)
Singapore	3.7 (1999)	1.7 (1999)
United Kingdom	4.2 (1998)	1.8 (1999)
United States	3.9 (1997)	2.9 (1999)

Source: Information Services Department (2002a).

population ratio were 5 : 1 beds per 1,000 population and 1 : 5 doctors per 1,000 population, respectively, again putting Hong Kong in a positive light compared with other developed countries (Information Services Department, 2002; see Table 2.3).

The main stress faced by the Hong Kong system has to do with financing. Since the 1970s, because of the rapid increase in population and structural changes in its epidemiological profile, advancement in medical technology and professional standards, as well as rising medical staff costs, the tax-funded public health care system had begun to experience strain on its capacity to cope with the expanding demand. Waiting lists started to lengthen, and public hospitals had to resort to using camp beds in the early 1980s. A lot of middle-class families gave up queuing and shifted to private hospitals. The disparity between occupancies in regional and nonregional hospitals, as well as between government and subvented hospitals, had deteriorated. Such imbalance and the problem of overcrowding gave rise to staff morale problems and public discontent. The staff of subvented hospitals also felt aggrieved about their less favorable fringe benefits and conditions of service compared with government hospital personnel, though both formed part of the public hospitals system. A former director of MHD, K.L. Thong, described such a system as a "colonial medical system" because of its benevolent nature, "always under-financed, under-staffed and certainly overcrowded, simply because it is not possible to provide medical services on such a universal scale without proper allocation of resources" (Thong, 1987, p. 32). A review of the provision of medical and health services was considered overdue. Several reviews took place during the 1980s and 1990s in an attempt to address the problems of supply.

Scott Report (1985)

In 1985, Australian consultant W.D. Scott was invited by the government to review the delivery of public hospital services. The Scott Report, published in December

1985, recommended drastic changes in the framework for the organization and delivery of public hospital care, resulting in the amalgamation of the then government hospitals sector and subvented hospitals sector into one unified network under a new statutory Hospital Authority, subsequently established in December 1990, to operate at arm's length from the government.

The traumatic reorganization process leading to the establishment of the HA, amidst strong opposition initially from public hospital staff unions and subsequent heated negotiation among the Provisional HA, staff unions, and the government over the terms of transition of civil service staff to HA, meant that little spare policy attention was given to other pertinent issues such as health finance. In fact, in order to make the HA (which by now had become an important piece of reform for the government) a sure success, the government not only had to concede to very favorable terms to public hospital staff, particularly medical doctors,[2] but also to inject extra funding to the new HA so as to enable a speedy improvement in facilities and quality of service. This resulted in a significant deviation from the original "cost-neutral" assumption of the Scott recommendations.

Public health care expenditure as a share of GDP had grown from 1.7 percent in 1989–1990 to 2.5 percent in 1996–1997, not to mention that during those years there was steady GDP growth as well. HA's staff increased in number from about 37,000 at its inception in 1991 to 51,203 in 1998–1999 (Hospital Authority, 2000, pp. 110–11). That period was in a sense the "golden age" of HA. The HA reform, however, had not addressed the supply-side problem, for 96 percent of the Authority's income still came from government revenue (Harvard Team, 1999, p. 26). Whether there had been productive efficiency gains in HA is somewhat ambiguous. According to the Harvard Team (1999, p. 42), who studied Hong Kong's health care system in 1998–1999, even though the operating cost per visit decreased for specialist outpatient service and accident and emergency (A&E) service between 1992–1993 and 1996–1997, the general outpatient service increased; for inpatient services, although the cost per discharge fell, the cost per bed per day increased. The same study also found that of the various input categories, the average growth rate during 1993–1998 was lowest for hospital beds (3.7 percent), but was much higher in staff (5.2 percent; especially at the consultant physician level, which was 13.5 percent) and capital expenditures, excluding the construction of new facilities (10.6 percent) (Harvard Team, 1999, p. 48, Table 4.15). Clearly, the rise in the cost of medical personnel and of medical technology and equipment accounted mainly for the increase in overall health care expenditures in the public sector.

Besides, there has also been a rise in demand. Because of its initial success and various service improvements, including the subsequent introduction of subsidized class B wards for middle-class patients, HA had managed to reverse the previous trend of flow of better-off patients to private hospitals. As a result, demand began to surge, thereby increasing its financial burden. Hospital discharge output and usage in specialist clinics and AandE services almost doubled in the decade from 1998–1999 (Hospital Authority, 1999a, p. 7).

Working Party on Primary Health Care (1990)

In 1990, a Working Party appointed by the government reviewed the delivery of primary health care and recommended the reorganization of public primary health care services to be administered eventually by a new Primary Health Care Authority (with the Department of Health as its executive arm). Closer integration with tertiary health care was also proposed, with the objective to achieve "*Health for All by Year 2000*"—then a motto of the World Health Organization (Working Party on Primary Care Health, 1990). The main thrust of the Working Party's Report was to argue the case for a primary health care–oriented health system. However, many of the report's recommendations, in particular the proposal to set up the Primary Health Care Authority, had been shelved by the government.

Neither the Scott Report nor the Primary Health Care Report addressed the problem of health finance in any significant way, both assuming that the government would continue to fund health care services through taxation revenue. The Scott Report did touch on issues of improved resource allocation, cost control, cost recovery and user charges (of an order of 10–15 percent), the introduction of less-subsidized higher class accommodation (i.e., in between nominally "free" general wards and the full-cost-recovered first-class wards), and even private health insurance schemes (Scott, 1985, chs. 10–11). The Primary Health Care Report accepted that it was imperative that a central policy on the future financing of health care services should be formulated and that "[t]he problems of how health care services should be funded, who should bear the cost and in what proportion should be addressed" (Working Party on Primary Health Care, 1990, para. 1.32). Yet, these issues were not actively followed up, and the public continued to be led to believe that with the establishment of HA and the resultant management reforms in public hospitals, more cost-effective and better quality medical services would be achievable.

Indeed, when Tung Chee-hwa took over as chief executive of the new Hong Kong Special Administrative Region in July 1997, he only named housing, education, and elderly welfare as his major policy priorities; health policy was not on his inaugural list (Tung 1997). Nobody had anticipated a crisis looming in the health sector as the HA reform continued to be portrayed as a success. By then HA had already become somewhat a victim of its own success. To start with, public expectations had been raised by the HA reform. HA also contributed to this through its initial improvements in service quality and medical technologies and facilities. If it continued to perform well, it would only attract more demand being shifted back from the private sector. However, in terms of capacity HA was not funded to cope with unlimited public demand.

"Towards Better Health" Document (1993)

Although the government's policy has all along been that no citizen should be denied adequate medical treatment for lack of economic means, the strains imposed by increasing demand, rising costs, and limited capacity to fund were becoming more obvious by the early 1990s. Between 1991 and 1992, the Health and Welfare Branch (HWB) convened two working groups—one to study medical insurance and the other to review fees and waivers. The findings and views of these two working groups were taken into consideration by HWB which produced a consultation document entitled *Towards Better Health*, published in July 1993, for public consultation.

The 1993 document recognized that the public health care system "has too narrow a financial base and there is not enough interface between public and private sector providers" (Health and Welfare Branch, 1993, para 3.3). It identified five options with a view to finding some acceptable means to make patients bear part of the health care expenditure:

- The percentage subsidy approach (implying that patients would bear a specified percentage of the cost of treatment in the form of user fees)
- The target group approach (comprising semiprivate rooms [i.e., Class B wards] in public hospitals, itemized charging, and target waiver groups)
- The coordinated voluntary insurance approach
- The compulsory comprehensive insurance approach
- The prioritization of treatment approach (somewhat inspired by the Oregon innovation in inviting the community to prioritize different types of treatment for public subsidy purpose)

Both the government and HA, with the backing of medical professionals who tended to think that the public was not sufficiently cost-conscious, were in favor of the percentage subsidy approach that would render public health care no longer a "free" service and thus discourage overuse. HA further advocated a "managed interface model" whereby HA services could be extended to cover some primary health care services within an integrated framework in collaboration with primary health providers. This was somewhat akin to the health maintenance organization (HMO) concept in the United States.

In presenting its options for reform, HWB had not addressed the question of affordability or how to enhance people's ability to pay for health care if fees were to be increased. Hence any suggestions aimed at generating more financial resources for health providers were bound to be one-sided considerations that could easily (and somewhat justifiably) be perceived as a climb-down on the government's long-standing commitment to universal health care. Strong public opposition and criticisms by political parties and some legislators eventually forced HWB to dump most of its main reform proposals. In the end, HWB concluded in January 1994 that only two options were pursuable—that is, to introduce semiprivate rooms in public hospitals with higher cost-recovery charges and to promote voluntary insurance. Itemized charging remained highly controversial and was put off the agenda. Despite a keen intention to revamp the existing regime of public health finance, the result of the 1993 review was to bring the situation back to square one. The government had lost both the argument and the opportunity for change.

1999 Harvard Report and Its Recommendations

The 1998–1999 Harvard consultancy, commissioned by the HWB[3] soon after the establishment of the new Hong Kong Special Administrative Region, was not only a revival of the 1993 exercise (though conducted within redefined policy parameters); it was also hastened by the increasing sign that the HA was proving to be a costly organization that would in due course become unsustainable financially unless new sources of income were to be found quickly.

The Harvard Team (1999) was asked to look at the strengths and weaknesses of the existing system of health finance and delivery to see if the system is sustainable in the long term, and if not, to propose alternative options. In its study, the team identified five objectives of reform as well as five options for improving the system. The objectives are:

• Maintaining and improving equity
• Improving quality and efficiency
• Improving financial sustainability by managing government budget on health and by better targeting of government subsidies
• Meeting the future needs of the population
• Managing overall health expenditure inflation

The five options are:

• Option A—status quo
• Option B—capping government budget on health
• Option C—raising user fees (so that by 2016 users have to bear 50–70 percent of medical cost with waivers continuing to apply to low-income groups on the government's social security assistance)
• Option D—a twin package of Health Security Plan (HSP) and a Long Term Care Savings Account (MEDISAGE) entailing employer and employee contributions, as well as government subsidy for the poor and unemployed
• Option E—competitive integrated health care, involving the full integration of both public and private providers competing within the framework of a central insurance system

The Harvard Team took a rather pessimistic view about the long-term financial sustainability of the current tax-funded public health care system. Assuming the current level of quality and access to public health services to be maintained and taking into account the aging population, increasing specialization in medicine, and the adoption of new (and costly) technology and rising public expectations, the team forecasted that in the next eighteen years, public health expenditure might take up 20–23 percent of the total government budget, a sharp rise from its current 14 percent (Harvard Team, 1999, p. 5). However, restrictions on public expenditure imposed by Hong Kong's Basic Law (i.e., the low tax and balanced budget requirements) meant that extra resources were unlikely to come about through additional taxation. The other major problem identified was that Hong Kong had a highly compartmentalized health care system that lacked coordination and cohesion between primary and inpatient care, acute and community medicine, and between the private and public sectors, thus threatening the organizational sustainability, quality, and efficiency of the system.

Evaluating the five options against the five objectives (see Table 2.4), the Harvard Team concluded that health finance reform should proceed according to Option D, then ultimately reach Option E. Option D would expect all citizens to enter into mandatory health insurance (HSP) funded by both employee and employer contributions, with the government providing a subsidy to pay the premium on behalf of the poor and unemployed. The insurance scheme would be managed by a central authority, Health Security Fund Inc., to pool the risks of the whole community and to serve as the informed purchaser of health services for the insured. Health Security Fund Inc. would pay a standard payment rate to any public or private provider that a patient might choose under the "money follows the patient" concept. Citizens would also have to contribute toward the MEDISAGE scheme, a compulsory savings account that would enable the contributor, upon retirement at age sixty-five or if incapacitated, to purchase a long-term care insurance package from private insurers. Option E would be built upon Option D, with the public and private health providers integrated through a competitive framework consisting of twelve to eighteen district-based "integrated health systems." This would mean the disintegration of HA, as well as the incorporation into an integrated regime of hitherto freely operating private general practitioners (GPs).

Reaction to Harvard Proposals

The proposals of the Harvard Team represented a drastic departure from the existing tax-financed and HA-delivered public hospital services system. It would also bring a great impact on private GPs who dominated some 70 percent of outpatient service business and who would face a central informed buyer of their service in the future in the form of Health Security Fund Inc. Both HA and private GPs, represented by the Hong Kong Medical Association, were against the proposed

Table 2.4

Evaluation of Options by Harvard Team (Compared to Existing System)

	Option A	Option B	Option C	Option D	Option E
Equity	Very good	Deteriorates	Deteriorates (reduced risk pooling)	Moderate improvement—risk pooling and patient choice	Moderate improvement—same as Option D
Quality	Variable	Deteriorates (because of reduced funding to cope with expanding demands)	Unchanged	Significant improvement ("money following patients" enables competition)	Significant improvement—same as and better than Option D; more integrated service systems and less compartmentalization.
Efficiency	Fair	Unchanged	Slight improvement	Significant improvement	Significant improvement
Financial sustainability	Poor	Significant improvement in terms of managing government budget for health; slight improvement in terms of better targeting government subsidies	Moderate improvement because of shift of financial burden from government budget to some patients; slight improvement in terms of better targeting of government subsidies (to the less well-off through exemptions)	Significant improvement with various built-in "control knobs" such as negotiated payment rates and demand side cost sharing; government subsidies to target elderly and lower-income households	Significant improvement—same as Option D; also benefits of integration

Meeting future needs of population	Poor	Same as option A	Same as option A	Moderate improvement—government resources can continue to target those least able to pay	Significant improvement
Managing overall expenditure	Fair	Same as Option A	Only slight improvement	Moderate improvement—through (a) separation of purchasing and provision; (b) negotiation on payment rates; (c) "money following patients"; (d) deductibles and co-payments to moderate consumer demand	Moderate improvement—same as Option D

Source: Adapted from Harvard Team (1999: 115, Table 1.2).

Option A—status quo;
Option B—capping government budget on health;
Option C—raising user fees;
Option D—HSP and MEDISAGE;
Option E—competitive integrated health care.

changes. The latter's worry was not lessened by the Harvard Team's criticism of some private GPs overcharging their patients.

Instead of going along the route of mandatory insurance and district-based competitive integrated care, HA counterproposed in its August 1999 submission to HWB a scheme whereby public hospital care would be prioritized into three levels of protection (Hospital Authority, 1999b). As the most essential service, Level 1 would be given full government subsidy from tax revenue. Level 2 services would be funded partially by tax subsidy and partially by patients' user payments. Level 3 services were to be strictly nonessential, the costs of which would be borne fully by patients who were expected to join a compulsory medical savings scheme and take out voluntary private insurance to ensure their ability to pay user fees. Instead of breaking up HA, the Authority advocated more interface with private providers through collaboration, coordination, contracting, and communication—similar to the managed interface model it first mooted in 1993. Put briefly, HA preferred keeping the status quo, with prioritized treatment and prioritized subsidy mechanisms introduced, and the supplementation of existing government tax funding by increased user fees. A comparison of the Harvard Team model and HA's alternative model is given in Figure 2.1.

The public reaction to the Harvard Team proposals was lukewarm and critical, to say the least. Major political parties such as the Democratic Party, the Democratic Alliance for the Betterment of Hong Kong (DAB), and the Liberal Party, as well as most legislators, were skeptical of the proposed health insurance and savings schemes, which they regarded as adding new hardships to the people.[4] Most press commentaries were against the proposals and highlighted instead the requirement for employees to shoulder more health care costs through insurance premiums on top of the tax they already paid, creating an additional disadvantage to the middle class (Hong Kong Policy Research Institute, 1999). Some academics questioned the Harvard Team's assumptions of rising cost and financial nonsustainability (Yuen, 1999), while others regarded the Harvard Team's proposed schemes as being too open-ended and not effective in capping overall health expenditures (Ho, 1999). Critics also pointed to the experience of Taiwan in setting up a national health insurance system in the early 1990s upon the same advice of William Hsiao who then headed the Harvard Team study of Hong Kong, which subsequently proved to be unduly expensive and resulted in increased premium payments from contributors in order to stay liquid. There was considerable concern that compulsory insurance might stimulate unconstrained demand (due to the moral hazard effect on patients) and lead to uncontrollable overall health spending. At present provider-induced patient demand is reasonably constrained by way of global budgets imposed by the government on HA, which in turn caps the operating budgets of individual hospitals.

A study completed by the Hong Kong Policy Research Institute in July 1999 found that most respondents preferred raising user fees to other options as a means to increase health finance and that public opinion was predominantly inclined to

Figure 2.1 Comparison of Harvard Team Model and Hospital Authority (HA) Model for Health Care Reform in Hong Kong

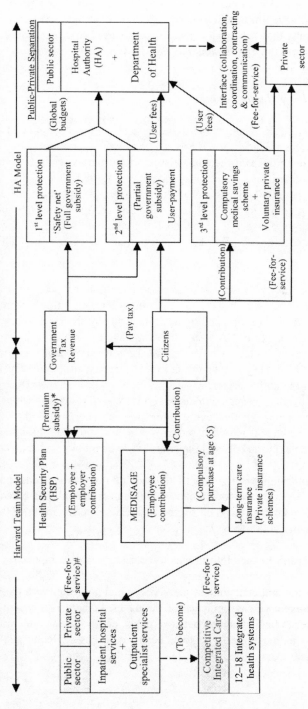

Notes: * Government subsidy for the poor and unemployed through premium; # Health Security Fund Inc. to pool risks of whole community and to serve as the informed purchaser of health services for the insured. It will pay a standard payment rate to any public or private provider that a patient chooses under "money follows the patient" concept.

retain the status quo. Views on core values in health finance seemed to convey a mixed sentiment, with almost all categories of response to the survey agreeing with both the statements that *"everyone should bear medical costs according to one's economic means"* and that *"everyone has the responsibility to save for one's medical expenses when in need, and should not wholly depend on the government."*

Toward Compulsory Health Savings

In September 1999, E.K. Yeoh, the HA chief executive, was appointed as secretary for health and welfare. Under him, HA was able to mount a reverse policy takeover of HWB. The Harvard Team recommendations were put aside. Instead, new proposals were announced in a consultation document entitled *Lifelong Investment in Health* in December 2000 (Health and Welfare Bureau, 2000), which was grounded in the thinking of the August 1999 HA submission. Advocating the concept of shared responsibility between the individual, the community, and the government, the consultation document proposed a reconfiguration of the existing public health care system:

- The Department of Health was to become mainly an advocate of health.
- The public sector was to take the lead in promoting family medicine as a means to improve health conditions while reducing the need to access more expensive tertiary care. HA would take on such a responsibility.
- The Department of Health was to transfer its general outpatient service to HA in order to facilitate the integration of the primary and secondary levels of care within the public sector.
- After such transfer, general outpatient service was to be redesigned into clinics primarily attending the financially vulnerable and chronically ill.
- Collaboration with the private sector was to be improved through adopting common clinical protocols and the joint development of new health care products.
- Government funding to HA was to be based on population needs and specific programs, reflecting the government's thinking on a population-based funding allocation model first unveiled by the financial secretary in his 2000 budget speech (Tsang, 2000, para. 25).

In addition, health finance reform was to consist of three components:

- Cost reduction and containment measures, which, together with the strengthening of preventive care, practice of family medicine, development of community care, and collaboration between public and private sectors, was to help to slow down the increase in total health care costs (Health and Welfare Bureau, 2000, para. 111).

- A revised fee structure was aimed at channeling more subsidies to lower income groups and the services of major financial risks to patients (Health and Welfare Bureau, 2000, paras. 113–14).
- Health Protection Accounts (HPAs) were to be established by requiring all working adults from the age of forty to sixty-four to contribute 1–2 percent of their earnings toward a compulsory savings scheme to meet their medical needs upon retirement. The savings would mainly be used to meet payments on public health care services, but the contributor could choose to use services in the private sector on the condition that reimbursements would only be made at the public sector rates, with the difference to be borne by personal means outside the savings account or from a separate voluntary insurance plan (Health and Welfare Bureau, 2000, paras. 118–21).

HPAs are similar in concept to what is being practiced by Singapore under its Medisave scheme. Enjoying as impressive achievement as Hong Kong in its health care system, with government operating 80 percent of the hospital beds, Singapore's health financing philosophy is very much based on individual responsibility. Patients there are expected to pay part of the cost of the medical service they use. Such a copayment principle applies even to the most heavily subsidized hospital wards. There are three tiers of funding: (1) the Medisave scheme—a personal account of the individual citizen within the Central Provident Fund, funded by 6–8 percent of employee salary, depending on age, out of total employer and employee contributions—helps the individual save for their health care expenses; (2) the Medishield scheme, the premium of which comes from Medisave funds, provides insurance coverage for catastrophic illnesses; and (3) the Medifund, which is an endowment fund set up by the Singapore government to help needy citizens to pay their medical bills (Ministry of Information and the Arts, Singapore, 2000, pp. 109, 112).

So far the idea of HPAs has not found much popularity in Hong Kong, as employees are already required under the newly effective mandatory provident fund legislation to contribute part of their wages towards a retirement provident fund (as from December 2000). Cross-party opposition to HPAs was voiced at the Panel on Health Services of the Legislative Council (2001). The economic slowdown triggered by the 1997 Asian financial crisis and now prolonged by the more recent global recession has made life harder for many income earners. To expect them to embrace another contributory scheme on health savings is politically contentious. The scope for increasing fees and charges would also be quite limited given the present heavily subsidized cost structure and the political sensitivity of any move on such an increase. Even if some fees were doubled, the additional fee income thus generated would still represent only a meager portion of total cost. Also the escalation in medical costs worldwide is partly caused by the rapid advances in medical technology, which facilitate speedier diagnosis and more effective curative treatment but certainly cost much more. Such technology-induced cost increase is difficult to contain because of society's continuous quest for new life-saving techniques and treatments.

Evaluation

The crux of Hong Kong's existing public health care system is that it is financially unviable to maintain a universal and almost "free" service. If universality and equity are considered primary social values, the means must be found to augment financial sustainability, either through additional taxation to generate more tax funding, or increasing user fees (whether by itemized charging or target subsidy) to supplement tax funding, or a fundamental overhaul of the system such as introducing a social health insurance scheme in place of tax funding. If the population finds user fee increase or compulsory health insurance unacceptable, then demand has to be managed. This can be achieved in two ways: either allowing health care providers to use their discretion to prioritize service and to focus resources and care on more needy patients, or redefining the existing government policy so that explicit target-subsidy would be adopted in place of universal coverage. This would result in some users, such as middle-class patients, being forced to turn to the private sector for more expensive care.

The latest HWB proposals in effect represent a combination of supply-and-demand-management strategies on top of the existing tax-funded public sector. On the supply side, user fees are to be increased to augment the system's income, as well as to achieve some kind of target subsidy. Citizens are also to be required to contribute to compulsory HPAs in order to boost up their ability to pay after retirement. On the demand side, HA is to take the lead in promoting family medicine and to take over general outpatient service so as to integrate both primary, secondary, and tertiary care within a single system (under HA) to make it more effective in gatekeeping patients from needing to use more expensive hospital services. Whether such an approach will work depends on how patients respond to the impact of fee increase and how the demand for hospital services can be scaled down as the impact of family medicine and better service integration is gradually felt.

The politics of economic recession presently prevailing in Hong Kong have made health finance reform all the more problematic. Up to the time of this writing, no further steps had been taken on the main proposals of the HWB consultation document, except the takeover of outpatient service by the HA and the introduction of a HK$100 charge for A&E service in November 2002, which was previously free.

Shanghai: Problems of "Societization"

Institutional Setting

Prior to recent reforms, the health care system in China was essentially made up of three components: (1) government-funded health care (*gongfei yiliao*) which covered state employees in government, military, and other state institutions (such as service units and schools) at all levels funded directly by the state budget; (2) labor insurance health care (*laobao yiliao*) which applied to employees in state-owned enterprises (SOEs) and most of the collectively owned enterprises that were financed by the enterprises' welfare fund; and (3) cooperative health care (*hezuo yiliao*) for the rural population, financed mainly by individual contributions and supplemented by collective funds. Hospitals were located mainly in the cities and major towns, and were owned and managed by either the Ministry of Health and its lower-tier health bureaus/departments or other state institutions such as the military, civil aviation, railway, and some large SOEs.

In Shanghai, before the founding of the People's Republic of China in 1949, most of the hospitals were either church sponsored or privately run. Government hospitals represented only 18 percent of the total number of hospitals and 37 percent of hospital beds (Shanghai Municipal Health Bureau, 1982, p. 9). With the transition into the socialist era in the 1950s, nongovernmental hospitals, along with virtually all forms of private enterprises and organizations, were nationalized. By 1956, all Shanghai hospitals were directly under state control. Private medical practitioners as a percentage of the total number of practitioners of both Western and Chinese medicine dropped from 89 percent in 1949 to 34 percent in 1955 and 7.4 percent by the mid-1960s. In 1980, there were only 173 private practitioners in the whole of Shanghai, representing 0.7 percent of the city's medical profession (Shanghai Municipal Health Bureau, 1982, p. 9).

From 1956 onward, the Shanghai municipal government had adopted a three-tier model in managing the health care system. The city's hospitals, clinics, and other health stations were grouped into three categories. Tier I providers (street-level hospitals, rural/county health clinics, and enterprise health stations) looked after the primary and preventive care of city residents, enterprise workers, and their families. Tier II providers (district, county, and enterprise-level hospitals) provided medical health care service within the relevant district, county, or enterprise. Tier III providers (municipal hospitals and teaching hospitals) handled complex treatments and health research and training. Each tier also performed a supervisory role over a lower tier.

The staff of government-run hospitals were state employees whereas those of enterprise hospitals were employees of the enterprise concerned. Most street-level hospitals and rural and township health clinics were categorized as collective units run on a self-accounting and self-financing basis; their staff were remunerated from fees collected and from government subsidies. Until recent reforms, all government cadres, staff of service units and educational institutions, as well as university students, enjoyed free medical care under the *gongfei yiliao* system. This system was jointly managed by the municipal Health Bureau and Finance Bureau and derived its funding from the government's budget in accordance with a per-capita average allocation. At the end of 1994, some 878,000 in Shanghai's population were covered by *gongfei yiliao*, involving a total annual expenditure of 555.64 million yuan, that is, a per capita average of 617 yuan (Project Team on Social Development Issues [hereafter Project Team], 1997, p. 85). Enterprise staff enjoyed free medical care under the *laobao yiliao* system while their families were half-subsidized by the enterprise. Individual enterprises managed and funded their own schemes of coverage in accordance with state policy directives.

Since 1958, the municipality's countryside had been covered by cooperative health care, with village production brigades and people's communes providing free or subsidized health care to the rural population. In 1970, there were over 2,700 villages that adopted cooperative schemes, under which participating villagers received reimbursement of a prescribed proportion of the cost of treatment and medicine, with the rest paid for out-of-pocket. By the 1980s, although rural cooperative health care had broken down in many parts of China, the system still remained much intact in Shanghai's countryside. According to some statistics, out of the 205 rural townships and 2,877 administrative villages under the city's jurisdiction, 95.69 percent maintained a cooperative health care scheme in 1997, covering 80 percent of the rural population (*Xinmin Wanbao*, 3 July 1996).

Achievements and Problems

China's medical and health care system had gone through difficult and traumatic periods over the three decades since the 1950s because of economic hardships, political turmoil, and erratic policies, particularly during the Cultural Revolution era (1966–1976). To overcome the low investment in urban health care institutions and the lack of any comprehensive health care protection schemes in the country, China had made extensive use of non-college-trained health workers and so-called rural doctors, formerly the "barefoot doctors." Despite various institutional and political constraints, health care achievements—in terms of life expectancy and the control and elimination of most endemic and epidemic diseases—were impressive. According to the minister of health (Chen, 1997), the infant mortality rate in 1997 was 31.4 per 1,000 live births, compared with 200 in pre-1949 days. The maternal mortality rate declined drastically from 1,500 per 100,000 live births to 61.9 in the same period. Life expectancy had more than doubled from only 32 in 1950 (World Bank, 1984) to 70.8 by 1997 nationally (Chen, 1997).

As a major metropolis, Shanghai enjoys an even more impressive achievement—with male and female life expectancy rates rising from 52.4 and 55.5, respectively, in 1952, to 76.38 and 80.53 in 1999 (Shanghai Municipal Statistics Bureau, 1993; 2000), comparable to the equivalent rates of Hong Kong. After some significant

Table 2.5

Number of Hospitals, Beds, and Doctors in Shanghai, 1949–1999

Year	Number of hospitals	Number of beds	Number of doctors	Bed-population ratio (per 1,000)	Doctor-population ratio (per 1,000)
1949	153	10,000	8,100	1.99	1.6
1950	135	10,800	6,300	2.20	1.3
1960	333	31,000	10,100	2.93	1.3
1970	341	39,000	18,100	3.64	1.8
1980	399	49,400	39,800	4.31	3.4
1985	405	53,200	48,500	4.4	4.0
1990	462	62,100	58,200	4.8	4.5
1995	485	66,900	53,700	5.1	4.1
1999	465	70,600	50,600	5.4	3.9
2000	459	73,070	49,872	5.7	3.8

Source: Shanghai Municipal Health Bureau (1982: 1–3, 13; 2000: 776). *Shanghai Municipal Statistics Bureau* (2001: 283). "Hospitals" include street hospitals with beds and village and county health institutions.

advances in the mid-1950s following a major overhaul of the city's health care organization, the supply of health care services had remained rather stagnant during the 1960s and 1970s, and many of the improvements there only took place in the early 1980s as China began to open up its economy and undertake various economic and institutional reforms (see Table 2.5).

Even so, bed-population and doctor-population ratios were at reasonable levels against the background of the adverse economic situation at the time, though such figures obviously did not speak much about the quality of service. Indeed, health care service remained largely constrained by the lack of funding. In line with socialist thinking during the early years of the new republic, medical fees were kept at a minimal level. Despite objection from the Ministry of Health, the central government promulgated three rounds of fee reductions in the late 1950s, resulting in fees contributing to only one-third to one-quarter of the costs (Peng, Cai, and Zhou, 1992, p. 5). Stringent state budgets coupled with unduly low fees meant that most hospitals and clinics were seriously underfunded. The situation was aggravated during the Cultural Revolution decade (1967–1976). As a result, the quality of care deteriorated and demands were grossly undersatisfied. Health staff were poorly paid and lacked motivation, while hospitals and clinics lacked funds to invest in new equipment and technology.

In Shanghai, hospital-bed utilization rates had gone down, from 85.2 percent, 93.6 percent, and 82.5 percent in 1956 for general hospitals, teaching hospitals,

Table 2.6

Hospital-Bed Utilization Rates in Shanghai, 1956–1978

Year	General hospitals	Teaching hospitals	Enterprise hospitals
1956	85.2%	93.6%	82.5%
1960	81.3%	86.8%	78.9%
1965	79.1%	91.4%	81.0%
1970	73.9%	89.8%	73.3%
1975	72.9%	85.7%	71.0%
1978	72.6%	84.0%	66.1%

Source: Shanghai Municipal Health Bureau (1982: 68, 75–76).

and enterprise hospitals, respectively, to 72.6 percent, 84.0 percent, and 66.1 percent in 1978 (see Table 2.6), reflecting worsening efficiency within the health care delivery system. Problems faced by residents were described as the "three hardships": "hard to see a doctor, hard to enter a hospital, and hard to have an operation."

Deregulation of User Fees since the 1980s

With the advent of economic reform in the 1980s, the health care management system had undergone rapid changes. In terms of health care delivery, both the supply and utilization of care had improved with the increase in hospital income through the "marketization" of health care. The number of hospital beds and doctors in Shanghai increased from 3.64 and 1.8, respectively, per 1,000 population in 1970 to 5.7 and 3.8 per 1,000 in 2000 (Table 2.5). Hospital-bed utilization rates moved up to 89.6 percent for general hospitals and 78.2 percent for enterprise hospitals by 1999 (Shanghai Municipal Health Bureau, 2000, p. 851), from 72.6 percent and 66.1 percent in 1978 (Table 2.6). Health care establishments had gradually moved from being units administered directly by the municipal Health Bureau into self-financing entities, with the principal source of income shifting from state budget allocation to fees collected from patients. Wages of health staff, hence their working morale, had improved, mainly through the distribution of bonus payments linked to the income of health care institutions.

Throughout China, in order to alleviate the funds-stricken hospitals and clinics in times of rising costs and inflation during the economic reform era, the state had allowed them to introduce some form of commercialization as a means to supplement income. User fees have since escalated, to the disadvantage of the urban poor and many rural households. Hospitals and clinics also sought to increase revenue through overprescription, inappropriate prescription and overcharging of drugs, as well as the performance of unnecessary but expensive procedures and surgical operations (Bo and Dong, 1993, p. 58; Sun, 1992, p. 60). A shift from the pre-1970s emphasis on preventive medicine to one more geared toward curative medicine and hospitalization has become apparent because of the changing financial incentive regime for health care providers, which encourages the consumption of more expensive hospital-based services under the new fee-for-service arrangements. Health care cost rising at an annual rate of over 20 percent was common in many Chinese cities during the 1980s (Zhou, 1995, pp. 3–15).

In Shanghai, the average per-visit outpatient fee increased from 1.89 yuan for county hospitals and 1.69 yuan for health clinics in 1980 to 3.89 yuan and 3.32 yuan, respectively, by 1987, while the corresponding increase in per-stay inpatient charges were from 70 yuan and 44 yuan to 198 yuan and 162 yuan—representing an annual increase of 10 percent and 42 percent (Cheng and Ye, 1982; Shanghai Municipal Health Bureau, 1987, pp. 158, 174). Medical and prescription charges continued to escalate throughout the 1990s (see Table 2.7). Per capita expenditure on medical fees and drugs for serving employees rose from 256 yuan in 1990 to 1061 yuan in 1997, and that for pensioners from 467 yuan to 1651 yuan (Shanghai Municipal Statistics Bureau, 1998, p. 59). Such rapid escalation in medical costs has aggravated the health finance problem.

Health Finance Reforms in the 1980s and 1990s

To tackle the problem of rising fees and overprescription, the Shanghai municipal government began to reform the *gongfei yiliao* system in the mid-1980s. After some pilot reforms at the district level, the municipal Health and Finance Bureaus introduced changes to all units at district and county township levels in 1988. Under these changes, health care expenditure was to be simultaneously linked to recipients, their work units, and providers so as to reduce moral hazard on their respective parts (Project Team on Social Development Issues, 1997, p. 86):

- Employees were given an amount of reserve payment for medical and prescription fees. When receiving treatment, they had to pay 10 percent of the outpatient charges and any sum up to 20 yuan. Charges in excess of 20 yuan would be borne by the work units. Pensioners and students were exempted from this requirement.
- Work units were allocated a fixed sum for self-management. Overexpenditure would only be covered by additional subsidy from the city government at its discretion.
- Medical payments (either outpatient charges or both such changes and inpatient fees) were placed under the custody of hospitals and clinics that had to account for them at the end of the year. Any surplus would be split (usually on 50 : 50 basis) between the government and the provider. In case of deficit,

Table 2.7

Health Care Service Quantity and Charges in Shanghai at County Level and Above, 1990–1996

	1990	1991	1992	1993	1994	1995	1996	1997	1999
Out-patient fee	15.4	19.8	27.5	44.5	58.3	79.2	88.1	100.2	117.2
Prescription									
(Chinese medicine)	15.0	16.5	23.4	28.8	34.9	48.3	56.3	62.5	66.5
(Western medicine)	12.2	16.5	22.8	34.3	41.6	54.0	62.4	66.8	73.2
In-patient charges	864	1,078	1,538	2,865	4,005	4,912	5,099	5,771	6,412
(of which, drugs)	(409)	(523)	(763)	(1,335)	(1,732)	(2,049)	(2,033)	(2,322)	(2,474)
Patients treated (in 1,000)	49,400	50,570	47,950	42,000	39,110	38,620	41,260	42,630	49,250
Patients discharged from hospitals (in 1,000)	623	616	627	610	598	596	639	682	773
Hospital-bed utilization rate	90.5%	91.6%	91.2%	86.4%	82.2%	84.6%	81.9%	83.8%	88.7%

Source: Shanghai Municipal Health Bureau (various years 1991–2000).

*All money values in yuan.

the government would only cover what was considered reasonable, with the remainder to be borne by the provider.

In 1996, pensioners were given a health care expenses supplement, and in turn were required to bear 5 percent of outpatient charges; as for inpatient expenses, serving employees had to bear 8 percent and pensioners 4 percent (Project Team, 1997, p. 87). Should such expenses go beyond the individual ability because of serious illness, the work unit could consider discretionary assistance. Units running into special difficulty could approach the finance authorities for help.

The enterprise *laobao yiliao* system was also reformed in the 1990s. Large SOEs implemented their own reform. For example, under Baoshan Steel's 1994 reform, workers were given monthly health care expense supplements (ranging from 30 to 90 yuan) but at the same time were required to pay for all of their hospitalization expenses and 20 percent of annual accumulated inpatient expenditures up to 10,000 yuan (and 2 percent of any amount exceeding 10,000 yuan); in case of serious illnesses such as cancer, 98 percent reimbursement would be provided for outpatient expenditure exceeding 2,000 yuan per year (Project Team, 1997, p. 87). In May 1996, the Shanghai municipal government promulgated temporary measures on hospitalization insurance for enterprise workers (Project Team, 1997, p. 87). Enterprises had to contribute 4.5 percent of the total last month's wages of their workers to a health insurance fund to provide worker hospitalization protection. In return, 85 percent of the worker's hospitalization expenses would be covered by the fund, with the balance "reasonably" shared between the worker and enterprise. Where hospitalization was caused by industrial injury or occupational illness, expenses exceeding prescribed limits were to be borne equally by the fund and enterprise.

In the rural townships, pilot reforms had been carried out since 1986. The essence was to introduce an insurance element into cooperative health care. Rural enterprises contributed 4 percent of their employed villagers' annual income, farming villagers 1.6–2 percent of average annual agricultural income, and the village collective economic units a per capita amount on behalf of the village population to a cooperative health insurance fund (Project Team, 1997, p. 88). In return, insured village families could claim reimbursement of various outpatient and hospitalization charges according to the government's prescribed limits.

All of these insurance-based reform initiatives in the urban and rural areas, which emphasized the responsibility of both individuals and their work units in bearing medical costs, were in line with nationwide health finance reforms that paved the way for eventually turning the health care protection system from one largely funded by the state and enterprise budgets into one supported by the individual, the work unit, and the community in the name of "societization" (*shehuihua*) of social services (Cheung, 2001; World Bank, 1997). Already there had been a clear shift in terms of health finance burden from the *gongfei yiliao*, *laobao yiliao*, rural cooperative health care, and semi-insurance schemes toward self-payment in the 1990s. For example, urban populations in major cities (including Shanghai) covered by *gongfei yiliao* and *laobao yiliao* had declined from 74.8 percent in 1993 to 52.3 percent in 1998, while self-paying patients rose from 16.4 percent to 34.3 percent (Ministry of Health, 1999, pp. 18, 27, 36).

Moving toward "Societization" of Health Care Protection

Health finance reform with an aim of "societizing" health care protection first took place in China in the form of pilot experiments in Zhenjiang and Jiujiang cities (the two *jiangs*) in late 1994. Following a National Working Conference on "Widening Pilot Spots for Employee Health Insurance System Reform" in April 1996, jointly convened by the State Economic System Reform Commission, Ministry of Finance, Ministry of Labor, and Ministry of Health, the State Council decreed that all provincial governments should introduce employee medical insurance reforms on the basis of the two-*jiang* experience, which was considered successful.

In essence, the two-*jiang* model reconfigured health expenditures into three tiers of financing. The first tier was the individual contributory savings account of the employee. The second tier consisted of out-of-pocket payments by the individual for medical fees that exceeded the balance in an employee's individual account. The final tier was the "social coordinating fund," which provided a kind of community risk-pooling device, to insure workers against the financial burden of catastrophic illness, beyond what could be financed by their individual account and out-of-pocket ability. Another feature of this model is the separation of purchaser and provider functions. Nowadays, there exist various local models which are variants of the original two-*jiang* approach (for more details see Zheng, 1997, pp. 336–44; also see Cheung, 2001). So far reforms have been more effective in shifting the financial burden to individual employees from the state and enterprises than in improving their ability to pay for the rising fees or in containing escalating medical costs.

In line with the national policy to establish individual health protection accounts and the social coordinating fund, on the basis of the reform experiments of the 1980s–1990s, the Shanghai municipal government promulgated Order No. 92— *The Shanghai Municipality's City and Township Employees' Basic Medical Security Measures* (Shanghai municipal government, www.sh.gov.cn accessed 21 November 2001)—in October 2000, which applies to all employees of urban and township enterprises, government organs, service units, social organizations, and privately run nonenterprise units. Under the new measures, all such employees have to contribute 2 percent of a "contribution base" calculated from their average monthly wages of the previous year toward the Medical Security Fund, while employer units contribute 10 percent. Retired employees are exempt from contribution. The Medical Security Fund consists of two components: a social coordination fund (SCF) and a personal health care account (PHA). Employees' contributions go into their respective PHAs, while employers' contributions almost exclusively go into the SCF, leaving only a minute proportion (0.5 percent to 1.5 percent, depending on the age of the employee) to the former's PHAs. For retirees, a higher proportion—from 4–4.5 percent depending on age—applies. On top of the municipal Medical Security Fund, lower-level authorities may impose supplementary contributions which go into local supplemental health protection funds.

The municipal Medical Security Bureau centrally manages the Medical Security Fund. It also sets the range of "insured" health care treatment, medical operations, and drug items, as well as the levels of payments in consultation with relevant authorities and in accordance with national regulations and guidelines. Insured employees receive medical treatment and get prescribed drugs from listed health care providers and drug dispensaries. Payments for outpatient treatment, AandE, and retailed drugs have to come from the employee's PHA. For major specialist outpatient treatment (such as radiotherapy and chemotherapy), the SCF bears 85 percent of the payment (92 percent in the case of retirees), with the rest funded by PHA. For hospitalization and AandE inpatient observation, SCF pays for 85 percent of the expenses above a payment threshold to be covered by PHA, calculated at 10 percent of the average municipal annual wage of the previous year. In all cases, any "personal" payments in excess of the accumulated balance of the PHA will have to be borne by the employee out-of-pocket. Lower threshold levels apply to retirees, with different rates depending on age. Furthermore, the SCF payment is capped at four times the municipality's average annual wage of the previous year.

Whether the reform in health care finance will ultimately succeed depends on two important factors:

- The ability to raise adequate funds to cover health care expenditure. Since the main purpose of health finance reform is to shift part of the financial burden from the state and enterprises to individual employees, it is necessary to ensure that the incomes of both parties are on the rise so as to ensure sustainability of their ability to contribute to the new insurance accounts and (in the case of employees) to make out-of-pocket payments in line with growth in health care expenditure. This is possible only in times of economic boom and where most enterprises are performing well and there is high employment. It also depends on the ability to keep health care costs under control.
- The effectiveness in cost containment. Health insurance schemes with appropriate individual and collective contributions will at best ensure coresponsibility to bear the consequences of health risk. They do not directly address provider-side problems such as escalating fee charges, unnecessary procedures and operations, overprescription, and overcharging on drugs. Unless there is a fundamental overhaul of the management of costs on the part of health care providers, particularly hospitals, health care cost will continue to escalate beyond what a reasonable steady growth in health insurance accounts can cope with.

Evaluation

While health finance reform seems to be inevitable, the transformation is not without pain and all kinds of constraints created by legacy and the shortage in ability to pay. There are several areas of concern:

- Limitation of coverage. Existing *gongfei yiliao* and *laobao yiliao* systems only cover staff of state organizations (government and service units, schools) and university students and employees of state-owned and collectively owned enterprises. These represent half of the urban and township population. Private enterprises, joint-venture enterprises, and the self-employed have yet to be fully covered by a comprehensive social health insurance scheme. Because of inadequate funding for rural cooperative health care, many poor villagers were unable to receive medical treatment for chronic and serious illnesses.
- Disparity in enterprise coverage. Different enterprises cover the health care expenses of their employees to different degrees—ranging from 100 percent for workers and 50 percent for their dependents of SOEs to 30 percent and nil, respectively, in the case of cooperative-type collective enterprises that participate in the health insurance fund operated by the municipal people's insurance corporation.
- Disparity between *gongfei yiliao* and *laobao yiliaio*. Many enterprises are financially constrained in providing adequate health insurance coverage for their workers, particularly those enterprises with a higher ratio of elderly workers and pensioners. According to the municipal Health Bureau's survey, staff medical expenditure as a proportion of total staff wages ranged from 2.08 percent to 510.87 percent in 1993 (Project Team, 1997, p. 89)!
- Escalating health care cost. The rapid escalation in medical cost (including the cost of drugs) since the 1980s has rendered the lowly funded health care insurance systems even more difficult to sustain financially. For example, in 1983 total health care expenditures of the whole of the city's workers under *laobao yiliiao* represented 7.35 percent of the total wage bill. By 1992, the ratio had jumped to 15.27 percent (Project Team, 1997, p. 89)—an annual increase of 29.9 percent during the decade, compared with only an annual growth of 5.57 percent between the thirty years from 1952 to 1982. As shown in Table 2.7, fees continued to rise rapidly in the 1990s. Average outpatient fees and hospitalization charges had grown 6.6 and 6.4 times, respectively, from 1990 to 1999. The charges on drugs formed 70–80 percent of outpatient fees and about half of hospitalization charges. Reform of the management of health care institutions is pertinent, in particular the delinking of health care providers' income from profits made in the administration of drugs.
- Impact of financial constraints. Financial constraints faced by many state and collective enterprises as well as the health insurance fund would result in them expecting workers to pay more for their medical bills out of contributory

individual accounts and out-of-pocket, hence reducing the impact of community-risk pooling. Those employees of funds-stricken enterprises and the urban and rural poor would become most vulnerable in the long run.

Comparative Overview

The health care systems of Hong Kong and Shanghai share one common feature, that is, the dominant role of the state in health finance. Shanghai, like the rest of China, enjoyed until recently a "nationalized" health care system where private practice was suppressed. Urban employees were protected under separate *gongfei yiliao* and *laobao yiliao* insurance schemes while the rural population depended on cooperative insurance support. User fees were very low. Service was affordable though of declining quality because of underfunding of the system, which then triggered calls for reforms. Nevertheless, the general health situation was satisfactory.

Similarly, in Hong Kong, while private GPs met much of the outpatient demand at market cost, the government was the predominant provider of hospital-based secondary, tertiary, and specialist services which were heavily subsidized by tax revenue, much like the U.K. National Health Service. User fees were very low, ensuring universal coverage and easy access. Health indicators of the population were impressive. Cost containment was generally effective while investment in health care staff, facilities, and technology continued at a reasonable pace. If not for the concern about the long-term financial sustainability of such a system under a tax-funding regime, the Hong Kong model may yet be cost-effective. For a comparison of the two models and the direction of recent reforms in both cities, see Table 2.8.

In their reform process, both Hong Kong and Shanghai see the need to allow user fees to be increased—as a source of additional income for health care providers and also to make individuals pay more for their health risk (hence rendering them more risk-conscious). Because of the serious underfunding of Shanghai's and other Chinese cities' health care systems, the pressure to increase fees is greater. Besides, the deregulation of medical fees since the 1980s in China and the expectation for health institutions to go marketized have led to these providers relying more and more on fees income. This has resulted in almost unrestrained provider-side moral hazard.

Shanghai's *societization* of health care is clearly a move to shift the financial burden from the state and enterprises to employees and workers. The prospect of such reform depends very much on an effective containment of medical expenditure, as well as economic growth that can improve the ability to pay of the working population. In the meantime, the reduction in insurance coverage has left many urban poor and less-protected rural families more vulnerable. The risks faced by Shanghai reformers are both economic and political. If the performance of the economy (and hence enterprises) becomes unsteady and there do not exist effective

Table 2.8

Comparison of Hong Kong and Shanghai Health Care Systems and Health Finance Reform Directions

	Hong Kong	Shanghai
Main features:	Tax-funded, budget-capping Very low user fees Non-integrated (between primary care and hospital services; between public and private sectors) Government health subsidy mainly in hospital service	Insurance-based system for state units and state-owned and collectively-owned enterprises; cooperative system for rural population Very low user fees (until 1980s) Integrated under state-run health care providers
Merits:	Universal coverage Equitable access and patient affordability (except in case of service provided by private GPs and hospitals where charging is not regulated) Health indicators impressive Cost-containment generally effective	Universal coverage (except countryside which depends on specific cooperative schemes) Equitable access and patient affordability (until 1980s) Health indicators impressive
Problems:	Budget limits implying problem of long-term sustainability	Under-funded—providers resorting to higher fees and income from drugs Cost-containment not effective
Reform direction:	Keep status quo Increase in user fees Compulsory health savings scheme to cover needs when retired Integration of government outpatient and hospital services and the promotion of family medicine in order to reduce use of expensive secondary care	Higher user fee and more self payment *Socielization* of health care by moving into compulsory contributory insurance scheme—with individual accounts and social coordinating fund Health care institutions to depend more on fee income than government budget

mechanisms to contain medical costs, any contributory health insurance system with significant user payment is bound to become financially unsustainable, resulting in either greater pressure for increasing contributions or deteriorating service coverage and quality. Opposition by the population who were used to more extensive coverage in the past should not be underestimated, especially when the current reforms of state-owned enterprises in China have resulted in massive redundancies.

Hong Kong's reform is more cautious and less traumatic. Despite the 1999 Harvard Team recommendations to switch from a tax-funded health finance system to one financed by compulsory contributory health insurance and savings schemes, the government and the wider community do not seem persuaded of the necessity in drastic reform. The current government proposals to combine fee increase with compulsory health savings accounts represent a more moderate reform package even though opposition to this scaled-down initiative is still significant. Even though the tax-funded health finance system will remain intact, it is clearly moving gradually from a universalist system toward one more based on target-subsidy of needier groups.

Notes

1. Enterprises, though paying directly for the health care and other welfare (such as housing and education) of their workers and dependents, as part of the "work unit" (*danwei*) system, had until recently been almost owned and operated by different levels of the state (Lu and Perry, 1997).
2. The final conversion package was that staff leaving the civil service to join HA could retain the existing civil service salary structure, but instead of civil service fringe benefits, the staff could be given a core benefits program in cash terms to cover retirement, death and disability, housing, leave, and medical protection. Such cash payment ranged from 13 percent of basic pay for junior nurses to as much as 65 percent for middle-rank doctors, consultants, and above (Provisional Hospital Authority, 1989, ch. 8).
3. Government policy "branches" have been renamed "bureaus" after the changeover of government.
4. Chan Yuen-han of the DAB and also a trade unionist, asked, "The Health Security Plan and the MEDISAGE proposal . . . are in my opinion fraught with problems of far-reaching implications. The report also added a suggestion on the so-called "deductibles." Can the people foot the bill?" Her colleague Choy So-yuk queried, "Under a collective contribution system, how can people be prevented from further abusing health care services on account of the health care insurance?" Sophie Leung, Liberal Party health spokesperson, bluntly stated that her party "does not think that this proposal is suited to Hong Kong's circumstances and needs." Even the Democratic Party, which appeared less hostile to the Harvard recommendations, had not found it possible to support them. Their health spokesperson Law Chi-kwong urged the government to move with great caution: "Next year [i.e., as from December 2000], the Mandatory Provident Fund Scheme will be implemented in Hong Kong and most people who do not have provident fund or pension protection have to make 5 percent contribution together with their employers from next year onwards, and they naturally oppose a medical insurance scheme with a stronger voice" (Legislative Council, 5 May 1999).

3

Housing

Kwok-yu Lau, James Lee, and Zhang Yongyue

Hong Kong as a capitalist economy has possibly one of the most overcrowded living environments in East Asia, with almost 7 million people being squeezed into high-rise living within a total land area of 1,098 square kilometers, where only 60 square kilometers of land[1] are for residential purpose. From a real estate development perspective, Hong Kong was, and to some extent still is, one of the world's most speculative housing markets where house price inflation reached an historical record of 21 percent appreciation within six months in the first half of 1997. Shanghai is now economically the most vibrant city in China, wielding almost the largest urban housing sector in East Asia. With a metropolitan population of 13 million, Shanghai effectively spearheaded the housing reform process through the selling of public rented housing to tenants, as well as providing Housing Provident Fund loans to encourage workers to purchase their own homes from either the state or the private housing sector. In a period of two decades, there have been enormous changes in the two housing systems, one severely damaged by the Asian Financial Crisis and still very much in the process of recuperation, and the other, much in the heyday of economic vibrancy, had everyone looking forward to joining the queue of homeowners. A passing comment from a German journalist traveling on an official tour of Shanghai in 2001 made the comparison of Shanghai more complex, "Hong Kong is a city of the past while Shanghai is a city of the future." Indeed, Shanghai's prominence as a city is not entirely legendary. Howe (1981) made it a focal point of intellectual analysis well in advance of Deng Xiaoping's announcement of China's modern economic reform.[2] Indeed, Shanghai is always the prism to appreciate change. Pye (1981) was right to suggest that the story of Shanghai provides for understanding the Chinese political process.[3] White (1981) further suggested that a fuller appreciation of the importance and success of Shanghai must begin from the suburbia, or more aptly, its entire vicinity—the Yangzi Delta region.[4] Following this line of thinking, what happened in the Shanghai housing system is nothing that could be understood simply through the spectacle of local interests. Today, Shanghai has successfully emerged from a housing system that comprises a tenure structure of 68 percent state rental housing in 1958, to more than 70 percent[5] private home ownership in 2001. This achievement is remarkable not only in numerical terms. It also carries a far greater impact on a proper understanding of the marketization process now still pervasively implemented in China. Originating from two entirely distinctive economic systems, Hong Kong and Shanghai are now sharing two common housing policy themes. First, both places show a clear trend toward a high rate of home ownership. Second, both have shown preference in the use of consumer housing subsidy rather than in "brick and mortar" housing subsidy. Third, both governments have come to regard housing policy as an important component of economic policy, with Shanghai more so than Hong Kong. This chapter is devoted to outlining the important policy change in the two housing systems. The main argument is that, while the two housing systems emanated from entirely different structural backgrounds, there are clear signs that in some ways they are converging. In this chapter we first provide a discussion on the context and housing conditions of each city, which is followed by an outline of the important policy changes and their impact on the housing system. A comparison of the two cities' housing policies is then presented with an attempt to draw out the similarities and differences and the challenges ahead for a sustainable housing system.

Shanghai

Like any other major Chinese city, residential provision in Shanghai in the four decades following 1949 was largely in the form of highly subsidized state rental housing distributed mainly through the work unit, or *danwei*. As the largest urban economy in China, Shanghai was one of the leaders in national housing investment and new house construction. From 1949 to 1990,[6] Shanghai invested 17.276 billion yuan, completed 65.82 million square meters of new dwelling space, and demolished 8.3 million square meters of substandard dwelling space. These had successfully raised the per person living standard from 3.9 square meters in 1950 to 6.6 square meters in 1990. Despite this improvement, housing was still an acute problem in Shanghai. At the end of 1990, 178,571 square meters of dwelling space were in dilapidated housing requiring demolition and redevelopment. In 1989, households with a living standard of less than 4 square meters per person still numbered 327,000, and 14.4 percent of the population was considered to be living under difficult housing situations. Shanghai's housing problem had reached a critical point in the late 1980s.[7] After over forty years of practice under the supply-oriented economy, the previous housing allocation system was found to be fault-ridden. From the perspective of housing development in Shanghai, the ten-year period (1991 to 2000) was unprecedented not only in terms of the huge housing stock produced, but also in terms of the transformations in the housing system due to market-oriented reforms. Further, the importance of this period lies not only in the progress it made in shaping the real estate market, but also in its influence on the creation of market economy laws and regulations for the well-being of the fast-growing housing market. In 1990, the State Council issued a new set of policies for the development of the Shanghai-Pudong area. In 1992, the Shanghai government initiated a two-level administrative system, under which the district governments have much more responsibility and power than before. In the same year, the Shanghai government also redefined both its role and its socioeconomic development strategies. The tertiary sector, including real estate, was given high priority in these new policy directions. The government set a standard average of 10 square meters of living area per person to be reached by the year 2000. Policy also stated that redevelopment of the 3.65 million square meters of substandard housing stocks was to be completed by the year 2000. The construction of major urban infrastructure has facilitated land development and redevelopment in Shanghai. Numerous demands and shortages of residents who have been relocated have resulted from infrastructure development and land leasing. Although in many cases, the real income of some residents has increased substantially, the quality of living conditions has not kept pace. Housing reforms in Shanghai have increasingly prompted more residents to become amenable to owning houses in the market economy context. As far as institutional structure is concerned, the early 1990s saw the inauguration of a number of important innovative measures, which carries far and wide impact in China's housing reform process.

Important Policy and Impact of Reform on the Shanghai Housing System

Establishment of the Housing Provident Fund (HPF)

Shanghai is the pioneer Chinese city in setting up the Housing Provident Fund (HPF) scheme which is modeled after the Singapore Central Provident Fund. It was first established in May 1991 as a compulsory saving scheme stipulated by the Shanghai City Housing Provident Fund Ordinance (April 1996). The current contribution rate from individual employee and employer is 7 percent each. From July 1997 onward, a noncompulsory supplementary HPF contribution was effected, and individual work units can determine the rate of supplementary contribution (from 1–8 percent). By the end of 2001, a total contribution from the employers and the employees amounted to about 47 billion yuan.[8]

Individual HPF account holders are eligible for a loan for home purchase if they have been contributing to the HPF accounts for a consecutive period of no less than six months immediately before their submission for a home-purchase loan. The HPF account holders must have joined the HPF scheme and made their respective contributions for an accumulated period of 24 months. The maximum loan amount per account holder is capped at 100,000 yuan for the past six years (or 130,000 yuan if there are supplementary HPF contributions) on the conditions that (1) the total loan amount is not higher than fifteen times that of account holders' accumulated sum of contributions and its interest earnings; and (2) the loan amount is smaller than 80 percent of the purchase price of the to-be-completed new flats, or 50 percent of the purchase price of completed flats (in the latter case the maximum repayment period should not be longer than ten years).

One year after the establishment of the Shanghai HPF scheme, loans were available and mostly used to provide loan support to work units for housing production. Statistics show that in the period 1991 to 1996, a large portion of the HPF loans were used in financing work units with HPF account holders or development companies to construct dwellings in the Comfortable Housing Project (*Anju Gongcheng*). Some loans were also provided for the improvement and reform of dilapidated housing and to build housing to relieve overcrowding problems.[9] Wu (1997, p. 55) notes that HPF loans paid for half of workers' housing capital projects.[10] By the end of March 2001, the accumulated amount of loans for this purpose was 11.971 billion yuan (Gao, 2001). In other words, 30 percent of HPF loans (40.093 billion yuan) were used to finance corporate organizations to construct dwellings. The financial difficulties encountered by some work units and development companies have caused delay in repayment of housing construction loans and hence a drainage of funds for prospective households.

HPF loans provided to support individual home purchase were only made available in May 1992. The proportion of the loan used to support individual home purchase was as low as 5 percent of all HPF loans in the early period. In other words, the majority of the loan was used to fund work units. Since 1999, provision of loan to work units on housing production had been terminated. By the end of 2000, HPF loans for individual home purchase constituted 95 percent of all HPF outstanding loans.[11]

At the end of 2001, the accumulated contribution to the Shanghai City Housing Provident Fund reached 47.061 billion yuan. A total of 36.293 billion yuan was used as home mortgage loans for account holders.[12] About 450,000 low- to middle-income families had made use of the loan from the Housing Provident Fund scheme for home purchase which resulted in a total purchase of 37 million square meters (constructed floor dwelling space). An estimate suggests that about 1.5 million people had benefited from the HPF loan (Huang and Jiang, 2001).

Despite its success in supporting individual home purchases, the Shanghai HPF scheme lacks sufficient funds. In the first half year of 1999, Shanghai HPF was short of over 2 billion yuan. Since 2000, the number of HPF account holders who wished to make use of the accumulated HPF savings for home purchase has been on the increase. In 2000, about 7 billion yuan were used to finance home mortgage and house construction projects. In the same year, there were only 8 billion yuan HPF contributions.[13] Take 2001 as an example when the amount of HPF contribution collection was 9.3 billion yuan. Of this amount 4.8 billion yuan were withdrawn upon account holders' retirement or used by account holders for home purchase. The remaining amount available for mortgage loan was only 4.5 billion yuan. According to official estimate, a shortfall of HPF home purchase loan

Table 3.1

Home Purchase Mortgage Loans Provided by Commercial Banks and HPF in Shanghai, 1998–2000

Year	Commercial banks' mortgage loans	HPF mortgage loans	Total mortgage loans	Percentage increase over previous year
1998	5.397	5.185	10.582	Not available
1999	10.90	7.353	18.253	72%
2000	17.76	8.645	26.405	45%

Source: Shanghai Housing and Land Resource Bureau (2001), "Taking the Opportunity to Further Develop Shanghai Housing and Real Estate Market: Analysis of Shanghai Housing and Real Estate Market," see www.cin.gov.cn/fdc/exp/2001032701.htm.

Unit: billion yuan.

amounted to 1 billion yuan in 2001.[14] The amount of contribution is too small and insufficient to meet other emergent needs of the HPF contributors. There is an urgent need to devise new ways for a viable HPF scheme to meet all needs.

However, increasing the rate of HPF contribution is not a feasible option because the total contribution rate of four compulsory schemes covering HPF, old age insurance, unemployment insurance, and medical insurance is already 42.5 percent for employers and 15 percent for employees (Cong, 2001). There is virtually no room for a further increase in HPF contribution. HPF contributors will lose confidence if the total amount HPF savings required to fund individual home purchase and to pay for those who retire, or to pay interest for account holders, or to pay for HPF administration charges, and so on exceeds the amount remaining on the HPF accounts.

Xie (2001b) suggested that the Shanghai Housing Provident has already started to develop the rudiments of a mortgage system to facilitate home purchase. As the amount of mortgage loan from HPF is limited, there is a need for home purchasers to obtain additional mortgage loans from commercial banks. Recent statistics show a growing trend in securing mortgage loans from both sources (see Table 3.1).

With these mortgage facilities, there is evidence that more people are induced to spend more on housing, which has become a major driving force of the Shanghai economy. By the end of March 2001, the total outstanding individual home purchase loan amount in Shanghai city was about 63 billion yuan. Out of this, 22.413 billion yuan (35 percent) were financed from Shanghai HPF (Wang and Jiang, 2001).

Officials of the Shanghai Housing Provident Fund Management Center have been conscious of the inherent problems of HPF despite its success in aiding individual account holders to purchase a home with a mortgage. In the earlier years of the HPF operation, less than 5 percent of the loans were used to support individual home purchase. From 1997 onward, the HPF loans were mainly used by individual account holders for home purchase. In 2001 individual home mortgage loans constituted 98 percent of all HPF loans.[15] The increased demand for this kind of loan is beyond the capacity of the HPF. The HPF Management Center has made adjustments regarding the down payment to house price ratio, allowing an extension of repayment period (from fifteen years to twenty years, starting from May 1998), a pegged mortgage loan ceiling to the amount of HPF contribution in the borrower's account, and so on. Unlike the commercial banks that can make a profit by charging borrowers commercial rates, the HPF policy of low-interest (pegged to commercial banks' saving rates) charge on borrowers and low-interest payment to account holders may have served the purpose of mutual help. But in the long run, it would not be financially viable for the HPF to meet all demand for low-interest home mortgage loans that are normally repayable in ten to fifteen years. To increase HPF loan capacity, the option of securization of the HPF loan portfolio has been under serious consideration by the Shanghai and central government authorities.

Sale of Public Housing

The Shanghai government promotes home ownership mainly through sales of public dwellings to the sitting tenants scheme. Between 1993 (pilot scheme) and 2001, 1.485 million public dwelling units were sold. As a result of this large-scale and cheap-price sale of public dwellings scheme, home ownership rate increased, and the government reduced its financial burden on subsidizing the deficit recurrent accounts of the public tenanted properties. In order to induce sitting tenants to purchase, the prices set for these public flats were rather low. Public tenants therefore would not need to spend all their savings to become home owners. By the end of September 1996, about 640,000 public flats (total construction floor space: 33 million square meters) were sold to sitting tenants.[16] By September 1999, 819,200 public flats (44.51 million square meters) were sold.[17] These constituted about 50 percent of all self-contained public flats identified as suitable for sale. By the end of 2000, the accumulated total number of public rental flats sold to tenants reached 1.364 million units or 71 million square meters of dwelling space. These constituted more than 75 percent of all public rental dwellings. During the same year, 559,800 public dwellings (29.7504 million square meters constructed dwelling space) were sold.[18] By the end of 2001, a total of 1.485 million public dwelling units (or 80.915 million square meters) were sold. These represented over 80 percent of all saleable public dwellings or 37 percent of all dwellings in Shanghai.[19]

Promotion of Resale Flats Transaction

There are four measures to promote sales of market dwellings and resale of public flats:[20]

1. Facilitating home purchase with easy access to home mortgage loan
2. Encouraging home purchasers to spend part of their savings
3. Allowing owners of public sales flats to enjoy capital gain upon resale of their flats in the housing market on condition that they trade up for market dwellings
4. Reduction of tax rates on home purchase and housing development projects

When the sale of public rental housing scheme was first promulgated, purchasers were told that they could sell their flats upon completion of payment of the sales price and occupancy in the purchased flats for five years. Receipts from resale have to be shared between work units (the original owners) and the owner. Surveys completed in 1996 in different districts of Shanghai city show that about one-third of the owners of public sale flats would wish to sell their flats to trade up for a better dwelling.[21] Measures to promote sales of market dwellings were developed against the context of a rising number of vacant dwellings. Vacant dwellings increased from 1994's 1.57 million square meters to 5.07 million square meters in 1996, and to 9.2 million square meters in 1999 (Shanghai Academy of Social Science, 2001).

A pilot scheme on resale of public sale flats was first tried out in three districts since August 1996. The scheme has been further extended to cover all public sale flats since July 1997. Owners of public flats with certificate of property ownership are allowed to sell their flats in the market at market prices and therefore would be able to obtain capital gain. The difference between the market sale prices and the cost prices paid previously is regarded as a form of compensation to the workers who have not been given a sufficient amount in their wages to pay for their housing. The capital gains would help ameliorate the affordability problem of high-market dwelling prices among low-income households in Shanghai. Between August 1996 and the end of 1999, 30,684 flats (1.563 million square meters) were resold. The total sales amounted to 3.449 billion yuan. Among them 90 percent used their capital gain to purchase another dwelling.[22]

Since mid-1998, public flats not suitable for sale have also been available for transfer of user's rights. Tenants of these flats are allowed to exchange use rights with other tenants or sell these public rental flats to other users/agencies to obtain capital gain and to trade up for purchase of a better dwelling. Between June 1998 and the end of 1999, 15,723 public rental units (596,400 square meters) were transacted. Of these 15,723 public rental dwellings, 3,460 were for exchange involving a total area of 158,600 square meters. The other 12,263 units (437,700 square meters) were for transfer involving a total transfer price of 902 million yuan. On average, each public tenant household sold 35.7 square meters of public rental space and gained 74,000 yuan.[23] Information shows that with such capital gain, households bought another flat (of average size 70.5 square meters) by paying 235,000 yuan. In other words, on top of the gain, they spent two times the amount gained in the transfer to purchase a better dwelling, doubling the size of the rental flat transferred. In 2000, 16,941 households registered their flats for transfer/exchange. Out of these 16,941 flats (682,000 square meters), 8,601 flats (282,000 square meters) were for transfer with a total transfer price of 620 million yuan; the remaining 8,340 flats (400,000 square meters) were for exchange between public tenants. On average each tenant household transferred 36.6 square meters and gained 72,100 yuan. Between June 1998 and the end of 2001, 71,062 public dwelling units (2.417 million square meters) were transacted. Of these 71,062 public rental dwellings, 21,324 were for exchange. The other 49,736 units were for transfer involving a value of 3,390 million yuan.[24]

Rent Reform and Low-Rent Housing in Shanghai

Rent increase has been conceived as a major component of housing reform in China. Rental charge largely doubled in 1990. Tenants were given a rent supplement at the rate of 2 percent of basic wages. In mid-1995, a 50 percent increase in rent was implemented. However, the actual rent was still low (at 0.75 yuan per square meter). Another 50 percent increase was added in October 1996.[25] On average, the public rental charge was 2.1 yuan per square meter of usable space. The average monthly household disposable income in Shanghai in 2000 was 2,743 yuan.[26] Assuming that average households live in a public dwelling of 50 square meters, the amount of the monthly rent charge was estimated at around 75 yuan, which is equivalent to a 2.73 percent average of each household's monthly income.[27] Another study shows that the ratio of annual per capita rental expenses to annual per capita actual income was only slightly increased from 1.13 : 100 in 1991 to 1.29 : 100 in 1999.[28] By the modern-day Shanghai standard, a monthly rental charge of 75 yuan is lower than the average household's spending on cigarettes, wine, tea, and drinks.[29] This suggests that rent reform is still a long way away from truly reflecting an efficient price.

As there is a plan to stop the in-kind housing allocation, the Shanghai government, like many governments of the large metropolis in mainland China, had also devised a plan to meet the need of the poor households. Provision of low-rent dwellings or rent allowance is seen as part of the safety net and is sometimes known as *shehui baozhang zhufang* (social security housing). The choice of Shanghai is to use a rent allowance to meet the housing needs of the low-income households (defined as those with a monthly income of less than 280 yuan per person) that are also living in an overcrowded environment (defined as those having no more than 5 square meters of dwelling space per person).

The Shanghai government prefers consumer housing subsidy (in terms of rent allowance) because it is believed that rent allowance recipients would have more choices in the private housing sector. When there are numerous vacant flats, any additional demand for private sector housing is welcome by real estate developers. A senior official of the Ministry of Construction[30] comments that rent allowance enjoys one great benefit—it can be withdrawn easily once the households are found ineligible. Moving ineligible households out of the cheap rental public dwellings is more difficult.

The scale of operation of the rent assistance to low-income households program is very small because the eligibility criteria are very stringent. A monthly per capita income of 280 yuan implies that they are the poorest among the poor. Information shows that the actual monthly income per person among the lowest 10 percent income group was 473 yuan in 1999 and 517 yuan in 2000.[31] At the end of 2001, 2,221 households submitted their applications for low-rent assistance. Of these, 1,610 households (73 percent) were verified as eligible. Among the 1,610 eligible households, 1,307 households (81 percent) had received a rent allowance or low-rent housing.[32] Less than 5 percent of eligible households took the in-kind housing allocation of low-rent housing.[33]

Hong Kong

In 1980, Hong Kong had a population of 5.06 million people. By 2001, it had increased to 6.71 million. That is a net increase of 1.65 million people (32 percent) over twenty years. In an area of 1,097 square kilometers, 83 percent of the space is non-built-up land.[34] About 5.3 percent is built-up residential area. Housing 6.7 million people in this small area makes Hong Kong one of the world's most densely populated urban environments. As a result of its hilly terrain, limited land supply, high land prices, and location preferences, the most intensive form of housing development in high-rise flatted blocks has to be adopted.[35] Consequently there is a requirement for huge capital costs in developing high-density and high-rise housing projects. The coexistence of a residual welfare system[36] and strong state housing makes Hong Kong rather unique in social policy terms. How could it be feasible? In 1999, 38 percent of households in permanent housing in Hong Kong were public rented; 14 percent of households purchased public sector home ownership flats; and only 12 percent of all households were in private renting. In other words, over half of Hong Kong's households were housed in the public sector rental and sale flats of reasonably good standard. The other half were in private sector housing which had diverse quality and prices. Over 75 percent of new private sector housing comes from seven major private developers.[37] In a densely populated city such as Hong Kong, the size of dwellings is normally very small. A government-commissioned survey shows that the average size of dwellings in Hong Kong is only 45 square meters.[38] In Table 3.2, the key phases of housing policy are highlighted.

Table 3.2

Key Phases of Housing Policy in Hong Kong

1. Before 1954: No provision of public rental housing
2. 1954–64: Meeting shelter needs of some low-income households
3. 1965–72: Development of self-contained social housing
4. 1973–87: Expansion of social housing program and redevelopment of sub-standard public rental estates—public sector supply–led strategy
5. 1988–97: Emergence of a state-managed housing system—promoting home ownership and private sector demand–led Strategy
6. Post-1997: Toward a stronger state-managed housing system—use of housing delimitation strategy

Source: Kwok-yu Lau, "Housing privatization: some observations on the disengagement and delimitation strategies," *Housing Express* (Hong Kong: Chartered Institute of Housing, November 2001), pp. 11–16.

Key Phases of Government Intervention in Housing

The Hong Kong government's involvement in the provision of rental housing is not planned. Its massive involvement in the public housing program was not started until the 1950s. The first resettlement blocks were built in 1954 to house victims of fires. These resettlement blocks were managed by the then Resettlement Department. The former Hong Kong Housing Authority completed its first social rented housing estates for white-collar workers in 1958. The quality of the early estates was of a low standard, composed of an H-shaped block with only communal toilet and washing facilities. In the mid 1960s, the government started to build low-cost public housing estates with improved housing standards for the eligible low-income households. Resettlement estates built after the mid-1960s were also improved in terms of dwelling size and internal facilities. From 1970 onward, the average living space per person had increased to 35 square feet per adult, and the average dwelling size increased to 238 square feet. Every family housed in newer design blocks could enjoy its own independent facilities, with toilet, kitchen, and water supply being provided within the dwelling unit. In the nineteen-year period between 1954 and 1973, 348,100 social rented dwellings were built. On average, 18,321 units were completed per annum.[39]

Although 1954 marked the beginning of the Hong Kong government's involvement in public housing, 1973 was a watershed year of planned expansion of the social housing program. Credit should be given to the then newly appointed Governor MacLehose. In October 1972, he announced a Ten-Year Housing Program aimed at providing 1.8 million people with adequate housing. The program did not expect the private sector to take much part and a supply-led strategy seemed to dominate housing policy. A new Hong Kong Housing Authority (HKHA) was established in 1973 to oversee and coordinate public housing development, taking over the Resettlement Department and the former Housing Authority. The introduction of Home Ownership Scheme (HOS; built-for-sales public flats) in 1977 was a clear departure from the traditional policy of only providing public rental housing. Despite this new breakthrough, the overall public rental and sales flats completed were only 55 percent of the original targets.[40] Another noticeable development during this ten-year period was the launch of the public housing redevelopment program that aimed at turning all substandard social rented dwellings into self-sufficient housing units of modern standards by conversion or in-site redevelopment. The massive development of New Towns also started in 1973, marking the beginning of an era of sub-urbanization.

As the demand for subsidized home ownership housing increased, the government took initiatives to develop a longer-term housing strategy. A significant policy change occurred in 1988. During that year a new scheme (Home Purchase Loan Scheme, HPLS) was set up to provide interest-free loans to eligible applicants to purchase private sector housing. This was because the government recognized the limitations and economic implications of expanding HOS. Private sector developers in general welcomed such a new initiative as it generated a new demand for private housing. Between April 1988 and March 2000, a total of 36,000 loans were granted to eligible applicants under the Home Purchase Loan Scheme.[41]

Taking note of the frustration and difficulties of middle-income households[42] to own affordable dwellings in a price-hike property market, the government further introduced the Sandwich Class Housing Scheme in 1993 and the Home Starter Loan Scheme[43] in April 1998. These schemes were administered by the Hong Kong Housing Society. A total of 5,700 families made use of the Sandwich Class Housing loan to purchase private dwellings in the six successive phases from 1993 to 1997. By March 2000, 10,446 Sandwich Class Housing Scheme flats were completed. Another 32,072 loans were drawn to purchase private dwellings under the Home Starter Loan Scheme by July 2002.[44]

The decision to sell about 25,000 public rented flats per annum to sitting tenants at heavily discounted prices was made in 1997, and the Tenants Purchase Scheme was successfully launched in 1998. Concern over the drastic drop of property prices after the 1997 Asian financial turmoil has also prompted some important changes in housing policy regarding the supply of Home Ownership Scheme flats. These are further discussed in later sections.

Impact of Policy Change on the Housing System

The HOS was the first public sector flats for sale scheme. The intention behind the policy was to provide better off tenants of rental estates with an opportunity for home ownership. The intended result was twofold: first to reduce the pressure on the rental sector and second to pave the way for a longer term privatization strategy. In addition, there was also the Private Sector Participation Scheme (PSPS). This was launched as a result of the Real Estate Developers' Association's expressed anxiety over the entry by government into the domestic flats for sale market. Under the PSPS, sites were sold to developers for the construction of HOS flats with stringent requirements. In return, developers were guaranteed a fixed price and assured profits. In a buoyant property market, the government actually made a handsome profit throughout the history of the HOS sale.

The expansion of the HOS was also based on pragmatic financial considerations. The method of price setting was ensured in the then prevailing market conditions, which paved the way for a fundamental shift in the financing of public housing in Hong Kong. HOS has been able to generate sufficient income for the Housing Authority to foot the loss incurred in the operation of the public rental sector.

The Asian financial crisis has fundamentally shifted the HOS policy. With a massive property market slump since 1997, the private housing market sale volume continued to contract until 2002. The declining world economy has aggravated the real estate market even further. Developers blamed the HOS policy as a factor continuing the slump. As the property prices continued to fall, the government made an announcement in September 2001 that it would impose a moratorium of HOS sales for ten months. At the end of the moratorium, the supply of HOS flats were down to around 2,000 flats per year. Such a drastic reduction in flats gave a signal to the developers that the HOS supply would not be the likely competitor of private sector housing. However, as the market continued to slump in September 2002, the developers escalated the demand to wiping out the entire HOS.

Home Purchase Loan Scheme and Other Loans to Assist Home Ownership

The introduction of consumer subsidies through interest-free home loans was the outcome of the 1987 Long Term Housing Strategy. "Choice" is the catchword for the introduction of many home loan schemes since 1988. The first was Home Purchase Loan Scheme (HPLS) introduced in 1988. HPLS is the least of the cost subsidies among the various government housing schemes (public rental housing, HOS/PSPS, and HPLS). Eligible applicants are offered an interest-free loan, re-payable up to a maximum of twenty years. Alternatively, they may opt for a monthly subsidy for forty-eight months, which need not be repaid. To increase its popularity, the loan and monthly subsidy were increased to $400,000 and $3,400, respectively. Tenant households in the private housing sector were eligible to apply if their total monthly income did not exceed $30,000 (income criteria for the period April 1997 to March 1998). To encourage more sitting public housing tenants to give up their rental flats, the loan and the monthly subsidy were further increased to $600,000 and $5,100, respectively. Such a loan could be used by public tenants to purchase HOS flats in the secondary market. By the end of March 1996, 12,137 loans and 651 monthly subsidies had been granted, and 7,121 public housing units had been recovered for reallocation.[45] Between 1988 and mid-2002, about 52,000

HPLS loans and monthly subsidies were granted to successful applicants.[46] Another loan scheme was called the Sandwich Class Housing Loan Scheme; it was administered by the Hong Kong Housing Society. It was first introduced in 1993 to help families with monthly incomes of between $25,001 and $50,000 buy their own home in the private housing market. The idea was to increase the threshold of assisted home ownership so that middle-class people who had problems of affordability could improve their access to home ownership. The loan scheme, with a government grant of $3.38 billion, was designed to provide assistance to 6,000 families. Successful applicants were able to borrow up to 25 percent of the flat price. Interest per annum was charged at 2 percent, which was much lower than the market interest rate (10 percent in 1997). By June 2000, some 5,700 loans had been granted.[47]

Since mid-1997 various new measures were introduced to help enhance the rate of home ownership. These were to help achieve the target of 70 percent home ownership rate by 2007 as pledged by the chief executive of the Hong Kong Special Administrative Region. It is an extremely ambitious target as the home ownership rate in 1997 was only 50 percent. The introduction of the Home Starter Loan Scheme (HSLS) in 1998 was one of these new measures. Under the scheme, low-interest loans were to be granted to first-time buyers. The income ceiling for eligible applicants was capped at $60,000 per household per month when the loan was first introduced. A sum of HK$18 billion was provided for 30,000 eligible families. Each successful family was allowed a maximum loan amount of HK$600,000. Interest was charged at the rate of 2 percent per annum and 3.5 percent for families with a monthly income of not more than $31,000 and for families with an income between $31,000 and $60,000, respectively. Application for HSLS was closed on 31 March 2002. By mid 2002, about 32,000 applicants were given approval to obtain loans in the Home Starter Loan Scheme.[48]

Tenants Purchase Scheme 1998

In 1997, the most significant housing privatization policy implemented was the Tenants Purchase Scheme (TPS). This policy offers an opportunity for at least 250,000 tenant households to purchase their own place at a discount over the next decade (1998–2007). This is seen as an important measure to promote home ownership among existing public housing tenants. It is packaged as an additional choice on top of the provision of the Home Ownership Scheme flats. The TPS has been rolled out in phases, and in the first four phases, about 25,000 flats per annum would be available for sitting tenants to purchase. Disposing of public rented flats in phases is intended to avoid the undesirable effect of a sizeable provision of cheap public flats to flood the market and consequently push prices of the property market. Nonetheless, there were criticisms arguing for a complete overhaul of such a scheme as it ultimately reduces sitting tenants' interest in the Home Ownership Scheme. The number of HOS applicants from sitting tenants dropped substantially from 45,000 in June 1997 to 5,117 in August 2002.[49] The Tenants Purchase Scheme has thus created an awkward internal competitor for the HOS.

Promotion of Resale Public Sold Flats with Relaxation of Resale Conditions

There have been resale restrictions in all public sales housing programs. Home Ownership Scheme flats and Tenants Purchase Scheme flats were subject to resale restrictions. Conditions of resale were introduced to minimize the chance of owners using subsidized sale flats for speculative sale. Before 1997, owners of HOS could (1) in the first five years from the first assignment sell back at original prices to the Housing Authority only; (2) in the sixth to tenth years from the first assignment, resale at market prices to the Housing Authority; and (3) after ten years, free transaction in the open market is subject to payment of a premium proportionate to the original discount price at sale. For the sake of creating an active secondhand market, restrictions on resale were relaxed in the late 1990s. Purchasers would be free to sell the properties in the open market after five years on the condition that the original portion of discount was paid to the HKHA. Purchasers could sell the properties to the Housing Authority within the first two years at the original price or to other public tenants and people applying for public rental flats at the agreed price between the vendors and the purchasers between year three and year five. Neither of these purchasers was required to pay the original portion of the discount to the Housing Authority. However, if the purchaser subsequently sold the property in the open market, that person was then liable to pay the original portion of discount on the prevailing sale price to the Housing Authority. After the purchaser had paid the original portion of the discount (the premium) to the Housing Authority, the property from then on could be transacted as freely as any other market housing.

Reform in the Public Rental Sector: Relating Subsidy to Housing Needs

A new Housing Subsidy Policy that required wealthy public tenants to declare income every alternate year after they had been staying for over ten years was put into effect in 1988. Public tenants are required to pay double rent if their household income exceeds twice[50] the prevailing Waiting List Income Limits. A further refinement of Housing Subsidy Policy was made in 1996, and double rent–paying households are required to declare their assets every alternate year after paying double rent for two years. Failing the assets test, tenants have to pay market rent if they continue to live in public rental housing. This policy was further revised in 1999. After paying market rent for one year, well-off tenants are required to vacate from public tenancy. These policies are grouped under safeguarding the rational allocation of public housing policy.

The policy requirements to declare income and assets are seen as measures driving out the well-off families and leaving behind only those very poor families in the public rental sector—the often-described "housing residualization" phenomenon. Empirical evidence shown in Lau's 1997 study rejects such a claim.[51] He argues that there has been a good social mix in public housing communities. The income and net asset limits screening out the well-off tenants have been generous. For example, in 2001/02, a four-person public tenant household was only required to pay one and one-half net rent plus rates if their monthly household income was between $34,525 and $51,788.[52] Only those four-person public tenant households earning a monthly income exceeding $51,788 and with net assets exceeding $1.42 million are now required to pay market net rent plus rates. According to the 2001 Population Census, among the four-person households in the nonowner occupier household category in subsidized and private housing and in all households in private temporary housing, a household earning a monthly income of $35,000 should be in the sixty-ninth percentile of these four-person households, and a household earning a monthly income of $50,000 should be in the eightieth percentile.[53] The fact is that four-person households in the nonowner occupier households category in subsidized and private housing and all households in private temporary housing earning a monthly income of $17,263 will not be eligible for applying for public rental housing. Permitting those four-person households earning at the level equivalent to the top twentieth income percentile in private sector housing to pay just one and one-half net rent plus rates is indeed generous. The good effect is that tenants in public rental housing are not all low-income households and that a reasonable social mix is found in the social housing sector. However, the claim that public housing rental subsidy is given in relation to housing needs is thus questionable.

Shanghai and Hong Kong: Convergence or Divergence?

Convergence

On a very broad level, the housing policies of the two housing systems converge in four areas:

1. Promoting home ownership
2. Housing commodification
3. Promoting the use of consumer housing subsidy
4. Fueling the growth machine, that is, housing policy as part of macroeconomic policy

Promoting Home Ownership

Both the Shanghai and Hong Kong governments have accorded priority in promoting home ownership. In Shanghai, a number of policies (sales of public dwellings to sitting tenants at low prices, the Housing Provident Fund, relaxation of resale conditions for public dwellings, and reduction of tax and other charges on property transactions) have given a combined effect to promote home ownership rate. In 1958, state rental housing accounted for 68 percent of Shanghai's tenure structure. In 2001, the estimated home ownership rate was 70 percent. By the end of 2001, over 1.485 million households had purchased public housing at a big discount. This, in effect, had contributed significantly to the increase of the home ownership rate in Shanghai in the past four decades. The sale of public dwellings to eligible tenants is the most significant public policy effected since 1993. Shanghai's sales of public dwellings program is well ahead of Hong Kong's. Table 3.3 provides an overall comparison of the Shanghai and Hong Kong sales of public dwelling schemes to eligible tenants.

The sale of public dwellings in Shanghai is part and parcel of a large-scale economic reform policy of Chinese cities. This is implemented at an unprecedented speed and scaled with the intention to create a privatized housing sector in which privatized dwellings will become a commodity for transaction in the market. However, in Hong Kong the degree of housing commodification is at a comparatively more advanced stage, and there is not such a fundamental shift. Therefore, when the TPS was implemented in 1998, the government adopted a very conscious attitude and was always wary of the impact on the housing market.

In Hong Kong, the HOS was designed to meet the home owning needs of those who could not afford private sector housing. In Shanghai, it was not until the adoption of the 1991 Shanghai Housing System Reform Program[54] that public dwellings newly completed or purchased from real estate developers were available for purchase. Comparatively speaking, Hong Kong started the sales of the public dwelling program at an earlier stage than its Shanghai counterpart. The combined effect of all home ownership promotion measures introduced since 1988 had pushed up the overall home ownership rate from 45 percent (1988/89) to 55 percent (2000/01) in Hong Kong.

Commodification of Housing

In Shanghai, public tenants have been provided with three attractive housing options: (1) to purchase the tenanted public dwelling with greatly discounted prices and then resell it for capital gain; (2) to stay on as a tenant but use rights can be exchanged with another tenant by moving to other public dwellings; and (3) to transfer the use rights to another family that is willing to pay a certain sum to obtain such rights.

The relaxation of the restrictions on the resale of public dwellings in Shanghai is consciously designed to fuel demand for private sector housing. In 1996, only 10 percent of the then occupied dwelling space was available for transaction in the property market. In 1999, it had been increased to 90 percent due to the relaxation of resale conditions and the new policy to allow public tenants to transfer and exchange use rights.[55] Those who sold their public dwellings usually bought another flat. On average each household has purchased an additional living space of 40 square meters. In September 2001, the cumulated number of resale public flats had reached 125,251, which is about 9.2 percent of all sold public rental flats.[56] Xie (2001b) also argues that the continued boom in the Shanghai property market is mainly due to the increased number of transactions of resale public flats in the secondhand market. Sale prices of resale public flats in 2000 are normally 25 percent cheaper than new private flats. It is anticipated that the liberalization of sale and purchase regulations will eventually vitalize the private housing market, since households with liquidity will seek better quality housing from the private market eventually. All in all the induced demand for residential housing in Shanghai has produced a multiplier effect on the economy in terms of activating other related sectors such as mortgage finance, property valuation and real estate agents, construction, fuel and other utilities, decoration, and furniture. The tax authority will also benefit ultimately.

In Hong Kong, the further liberalization of the resale conditions and the creation of the Secondary Market Scheme as well as the increased use of home purchase loans to replace HOS are obvious evidence of a more advanced stage in the process of housing commodification. HOS and TPS transaction records in the Secondary Market Scheme (SMS) suggest that the number of resale public dwellings is quite limited. Between mid-1997 (when SMS was first launched) and late 2001, a total of 11,931 flats was transacted. No follow-up survey has been conducted to track the next housing move of those who have sold the public flat. In this regard, Shanghai has taken a more thorough approach in housing commodification.

Promoting Use of Consumer Housing Subsidy to Boost Housing Demand

Evidence shows that there has been increasing use of consumer housing subsidy in both cities. At the end of 1998, the People's Congress in Shanghai laid down the basic principle to guide "housing monetarization"—the use of cash assistance to replace housing in-kind allocation. Direct cash subsidy to assist home ownership was to be provided to serving employees who had not yet been allocated welfare housing. The policy also added a cash supplement onto wages for the newly employed staff and employees to help them meet the costs of market housing.[57] In Shanghai, purchasers of public dwellings were encouraged to trade up. Poor households were also given rent allowance to rent private sector flats. There were also policy decisions on renovating and redeveloping the substandard housing (known as the 365 project as it involved demolishing 3,650,000 square meters of substandard space in the ten years ending in 2000), and households affected by the clearance projects were given monetary compensation and a special discount when they purchased dwellings in the market.

In Hong Kong, there are interest-free or low-interest loans to help prospective

Table 3.3

A Comparison of the Programs on Sales of Public Dwellings to Tenants in Shanghai and Hong Kong

	Shanghai	Hong Kong
Year of Commencement	1993	1998
Extent of Coverage	All public dwellings except those designated for redevelopment	To be implemented by phases in a 10-year period, about 25,000 flats were available for sale each year
Pricing	Below cost, minimum prices at about a quarter of cost prices	Based on adjusted replacement cost approach plus further discount with special credit arrangements (e.g., purchasers paid 12% of the assessed market value to obtain a public dwelling worthy of 30% of assessed market value if purchased in the first year of sale)
Resale Price	Upon payment of the difference between cost price and original payment, capital gains generated from resale can be used to pay for purchase of improved-quality dwelling	Payment of the original portion of the discount (70%, also known as the premium) if property is sold in the open market in the 6th year or in subsequent years. Between the 3rd and 5th years, the responsibility of paying the above premium will be passed on to the vendor as the public dwelling can only be sold in a restricted secondary market (mostly sitting public tenants)
Target of Resale Flats	No restriction	No restriction from 6th year onwards, but restricted to green form certificate holders between 3rd and 5th years or in subsequent years
Percentage of Public Dwellings Sold	80% of all saleable public dwellings	About 70% in the first four phases (between 1998 and 2001)

(continued)

Table 3.3 *(continued)*

	Shanghai	Hong Kong
Number of Public Dwellings Sold since Its Com-mencement (See note)	1.485 million units (1993 to 2001)	70,000 units (1998 to 2001)

Note: Figures in Shanghai included public dwellings built for sale since their first completion. Figures in Hong Kong only included those public rental dwellings selected for sales under the Hong Kong Housing Authority Tenants Purchase Scheme.

owners gain access to home ownership. The number of loans increased in the past few years as supply of HOS flats decreased (see Table 3.4).

The plan to use rental allowance (in cash) to serve as an alternative means to providing public rental housing has also been actively considered by the Hong Kong Housing Authority. A piloted rent allowance scheme is already in place for eligible elderly tenants. By the end of March 2002, only 122 eligible tenants had opted for the rent allowance despite a quota of 500 which was set in the pilot stage of the Rent Allowance for Elderly Scheme.[58] The lukewarm response to the elderly rent allowance is clear evidence that the cash allowance is not a complete substitute in meeting the different housing needs of specific groups of people.

Fueling the Growth Machine: Housing Policy as Major Macroeconomic Policy

Housing and real estate industry is considered to be one of the six main industries of the economy. Housing reform is considered the growth machine, and real estate is a major driver of Shanghai's economy. In Shanghai the contribution of real estate to GDP was about 2 percent in 1994. It rose to 5 percent in 1998 and 6 percent in 2001.[59] It is forecasted that the contribution percentage will rise to 11–12 percent by 2010.[60] Housing policy is construed as one of the major macroeconomic policies that stimulates and induces internal consumption. The induced demand for housing and related consumption is expected to boost supply. Statistics show that the pro-portion of fixed asset investment in residential housing to Shanghai city's fixed asset investment increased at a high rate. In 1978, it was only 7.4 percent, while ten years later, it rose to 14.2 percent and to 29.8 percent in 1994. In 2000, fixed asset investment in residential housing constituted 25.1 percent of Shanghai city's fixed asset investment.[61] The launch of the sale of flats to sitting tenants scheme in 1993 on the one hand is a definancing measure (to reduce the financial subsidy arising from managing public rental properties) and on the other hand a strategy of promoting the use of savings for home purchase.

In Hong Kong, the property market is always a key pillar of the economy: the

Table 3.4

Number of New HOS/PSPS Flats for Sale and Number of Loans Available in HPLS

Year	Number of new HOS/PSPS flats available for sale	Number of home purchase loans available for borrowing
1997/98	31,018	4,500
1998/99	12,427	10,000
1999/2000	14,510	4,500
2000/01	10,293	7,000
2001/02	6,307	16,500[1]
2002/03	2,451	10,000[2]

Source: Hong Kong Housing Authority, February and September 2002.

Notes:
1. Loan quota earmarked for the period April 2001 to June 2002
2. Loan quota under a new scheme—Home Assistance Loan Scheme to take effect in late 2002.

contribution of real estate and related sectors to GDP was about 21 to 30 percent between 1980 and 1998.[62] The percentage of land revenues and other property-related revenues (including general rates, property tax, stamp duty, and profit tax arising from property) was 54 percent in 1997/98 which has since dropped to 26 percent in 2001/02 when the property bubble burst. The finance market is also very much dependent on the prosperity of the property market. The ratio of building and property development corporate loan and individual home mortgage loan to total amount of loans was as high as 51.4 percent in 1998, and in 1992, it was at 40.5 percent. Due to a 60 percent drop of property prices (the drop from its peak at fourth quarter 1997 to fourth quarter 2001), the residential property prices in 2002 went back to the 1992 level. The Monetary Authority estimated that there were 81,000 property owners on mortgage suffering from owning negative equity.[63] The remaining 1 million homeowners of residential housing suffered from enormous devaluation of their properties. Moreover, the price fall also created a chain effect on the business sector. In the past many small businesses had used paid-up property for additional finance to support business operation. The continued fall of property prices had put many small businessmen into financial difficulty. Given the importance of housing in the economy, it is understandable that the Hong Kong government used housing policy as part and parcel of the macroeconomic policy— that is, the reduction of the public HOS flat supply and the increase of home purchase loans. Put simply, the housing sector is always one of the most crucial independent variables in the whole economic equation. Elsewhere, it has also been argued that East Asian economies have a unique tendency of using real estate and the financialization process as a major source of growth (Smart and Lee, 2003, forthcoming).

Divergence

The housing policy reforms of the two housing systems diverge in two areas: housing finance and housing provision for low-income households.

Housing Finance

In Shanghai, most home purchasers of public dwellings pay outright as prices of public flats are generally cheap. However, as property prices of both newly completed public and private dwellings go up, it is more difficult to purchase only with savings. For those who have not had adequate finance, they would then need to get help from the finance market. From mid-1992 onward, account holders of the Housing Provident Fund could make use of its provision for home purchase while the remaining part was financed by commercial bank mortgage loan. As the amount of loan from HPF was capped, there was a need for many home purchasers to obtain additional mortgage loans from commercial banks. Statistics of 1998, 1999, and 2000 show a growing trend in securing mortgage loans from both sources.

The provision of HPF home mortgage loans has in effect induced new requirements for mortgage loans from commercial banks. With these mortgage facilities, there is evidence that more people are induced to spend more on housing.

In contrast, Hong Kong home purchasers normally rely on banks and finance institutions to arrange home mortgage as there is already a very mature financial market. Furthermore, the Hong Kong Mortgage Corporation Limited (HKMC) was established in 1997 to promote the development of the secondary mortgage market. This new institution will help to address the problem of funding risks, namely, maturity mismatch, the liquidity risk, and the concentration risk.[64] In mid-2002, the HKMC implemented a Home Owner Mortgage Enhancement (HOME) Program. This program provides mortgage insurance to cover a bank's credit exposure above 90 percent and up to 140 percent of the current value of the property at the time of refinancing. The HOME Program aims to alleviate the financial burden of homeowners in negative equity and to provide an effective tool for banks to reduce the credit risk of mortgage loans with the current loan-to-value ratio above 100 percent. The successful operation of the HKMC will in the long run contribute to banking stability, debt market development, and strengthening home financing in Hong Kong.

There are also other incentives to induce potential home purchasers to buy. The Hong Kong Housing Authority provided interest-free home purchase loans to assist eligible applicants to purchase resale public dwellings. Charges are set at below market interest rate for those middle-income households borrowing from the Sandwich Class Housing Loan Scheme or the Home Starter Loan Scheme. In this regard,

Hong Kong appears to be more generous than Shanghai as successful applicants of these assisted loan schemes are not required to contribute to the compulsory saving schemes such as the Shanghai Housing Provident Fund, yet they are given interest-free or below-market interest charges for home purchase.

Housing Provision for Low-Income Households

Both governments devise stringent eligibility criteria for housing assistance. In Shanghai, only the poorest among the poor are eligible for rent assistance or low-rent dwellings. In 2000, the actual monthly income per person among the lowest 10 percent income group in Shanghai was 517 yuan. To be eligible, the poorest households should not have a monthly income exceeding 280 yuan. This to some extent explains why Shanghai only had 2,221 households applying for rent assistance at the end of 2001. Moreover, under the Chinese household registration system (hukou system), a poor migrant family without a Shanghai hukou is also not qualified for housing assistance.

Despite its claim of being a capitalist society, Hong Kong's social housing provision is enviable by Shanghai's standard. The eligibility criteria are less stringent. In recent years, about one-third to 40 percent of households in the private sector nonowner-occupier sector and in the temporary housing in Hong Kong are eligible for public rental housing. Moreover, members of migrant families, once joining their relatives in Hong Kong, are qualified for public rental housing. As of April 2001, a total of 108,433 households registered on the public rental housing waiting list.

Conclusions

Tenure Shift

The most noticeable change in the Shanghai housing system is tenure. Home owning has already become the dominant tenure. The various housing reform and commodification measures have successfully made half of the household home owners (Lu, 2001, p. 12). As more people become property owners, the residential areas have become people's space for private life. This is in contrast with the former notion of residential community for the labor force—a hinterland of the production base where employees were housed to facilitate production. Lu (2001) points out that private property rights bring about owners' concern and participation in community-building.

In Hong Kong, home ownership policy has increased home ownership rates from 45 percent in 1989 to 55 percent in 2001. Through various subsidized schemes, some 379,000[65] public and private tenant households have become owners since 1988. Based on these statistics, the Hong Kong government can claim its success in the home ownership promoting policy. However, beneath this success story lies the ever-increasing number of negative property owners, many of whom were induced to become home owners as a result of the government's active home ownership promotion. As the property bubble burst in 1997, the government was set to reconsider the risks involved in a property-owning economy. Many aggrieved property owners put the blame on the government as they argued that had there been no inducement many could have stayed as public housing tenants or opted for private renting. The chief executive of the Hong Kong SAR stated clearly in 1997 that "owning one's home is an aspiration shared by the people of Hong Kong. . . . The aim is to achieve a home ownership rate of 70 percent in ten years."[66] This was again reiterated in the 1998 *White Paper*: "the Government is fully determined to achieve this target, which will foster social stability and a sense of belonging, and help families to provide for their own future financial security."[67] However, the experience of many home owners in Hong Kong after the 1997 Asian financial crisis proved that such a "belief" is unfounded. The empirical evidence found in the last five years (1997 to 2002) suggests the contrary. Coupled with the rising unemployment rate (from 2.2 percent in 1997 to 7.7 percent in the second quarter 2002) and the threat of salary cuts, the lackluster property market persists despite an eleven-time reduction in interest rates to boost confidence in 2001. The increased number of discontented owners will become the center of social instability as pessimistic sentiments continue while leading bank economists forecast a gloomy economic prospect for the next five years.[68]

Forrest and Williams (1997, p. 202) have rightly pointed out that national economies are now more open and vulnerable to global economic forces. The Asian financial turmoil has delivered a lasting impact on the economy. There are a high number of job losses and reduced earnings. These in turn have a direct bearing on the two domestic housing markets. Unfortunately neither government of these two cities nor the people in the home-owning sector have understood the risks associated with home ownership in a volatile global economy. To sustain a home owning community or to promote a higher home ownership rate through housing privatization policies, government would need to develop policy instruments to help home owners manage their increased risks. The proposal by Holman (1997) on setting up a Mortgage Assistance Fund to help home owners as a consequence of a fall in income should be seriously considered by policy makers.

General Improvement of Living Conditions

Among all major cities in China, Shanghai is best known for its poor housing conditions. The per capita living space among urban dwellers was extremely low from 1950 to 1968 (3.9 square meters in 1950 and 4.0 square meters in 1968). Capital investment in residential housing projects was very limited between 1951 and 1977. Within these twenty-seven years, the total capital investment in dwellings was only 1.068 billion yuan. The capital investment in the following four years had already exceeded the total investment of the previous twenty-seven years. Table

3.5 shows that in the Sixth Five-Year Plan Period (1981–1985), fixed asset investment in residential housing jumped to 3.965 billion yuan, which doubled the total amount of housing investment of the previous 30 years (see Table 3.5). The rate of increase in investment in residential housing is particularly impressive in the 1990s. As a consequence, per capita living space was raised to 4.5 square meters in 1978. From then on there has been a steady increase, and it was further increased to 6.3 square meters in 1988 and 6.6 square meters in 1990.[69] Prior to the launch of housing reform in the early 1990s, the Shanghai municipal government regarded housing the "No.1" social problem. Even in the early 1990s, a sizeable proportion of households still could not afford living in self-contained housing units: 64.5 percent had to share bathroom facilities with other households, 55.8 percent shared toilets, 32 percent shared kitchens, and so on.[70] After years of government effort in improving general housing conditions, the average living space (sometimes also translated as housing space or dwelling space) per person was raised from 8 square meters in 1995 to 10.9 square meters in 1999. It was further increased to 11.8 square meters in 2000.[71] Moreover, the proportion of self-contained dwellings has also increased from 44.5 percent in 1995[72] to 85.7 percent in 2001.[73]

Despite these changes, the Shanghai government still cannot afford to be complacent about its achievements. A survey completed in 1998 shows that among twelve indices on life quality, housing condition is listed as the one with which Shanghai people expressed the least satisfaction. Another survey completed in April 1999 shows that among the 5,000 randomly selected respondents, 47.9 percent and 13.1 percent, respectively, indicated that their housing conditions were unsatisfactory and very unsatisfactory.[74] In other words, 61 percent of the respondents were not satisfied with their housing. The size of the dwelling units could be one source of dissatisfaction. The same survey confirms that 44.8 percent of households surveyed lived in a dwelling of 36 to 54 square meters, which is normally referred to as a two-room dwelling unit. Another 38 percent of households were accommodated in flats of 18 to 36 square meters, which is normally referred to as a one-room dwelling unit. Moreover, the indicator of dwelling space per person has to be read with caution. In Shanghai, the official statistics on dwelling space per person are computed by dividing the total dwelling space occupied by both residents with official resident status (*hukou*) and those without official resident status (including the floating population) by the total number of residents with official resident status. The total floating population as of April 2001 numbered 3.8 million, which was 23.1 percent of the whole population (16.7 million) in Shanghai Municipality.

In Hong Kong, three in ten households in 1985 were classified as inadequately housed. In 2001, only one in twenty households belonged to this category. In the public rental housing sector, about one in ten households had a living density of less than 4.5 square meters per person in March 1988. The massive housing construction and redevelopment programs have helped to improve the general living conditions. In March 2001, less than one in one hundred households had a living

Table 3.5

Shanghai City Fixed Asset Investment in Residential Housing and Dwelling Space Completed

Period	Fixed asset investment in residential housing (unit: 100 million yuan)	Dwelling space completed (unit: 10,000 square meters)
1951–1977	10.68	1,582 (1952–1977)
1978–1980	8.26	719
1981–1985	39.65	2,026
1986–1990	113.3	2,245
1991–1995	843.4	3,780
1996–2000	2,040.2	7,809

Source: Shanghai Statistical Yearbook 2001, Table 20.56. (Internet Version http://202.109. 72.30/tjnj/2001/tables/20_56.htm.)

density of less than 4.5 square meters per person. The median living density of Housing Authority public tenants increased from 6.2 square meters per person (internal floor area) in March 1988 to 9.6 square meters per person (internal floor area) in March 2001.[75]

Future Housing Challenges

While recognizing the important contribution of the residential housing sector to the overall growth of Shanghai's economy, we must be cautious that the rapid increase in fixed asset investment in residential housing will place too much strain on the capital finance market. Such caution is based on the large number of vacant dwellings (see Table 3.6). An estimate suggests that 22.65 billion yuan were locked due to 9.016 million square meters of vacant market dwellings.[76] The vacancy rate reached a high level which merited careful attention of the relevant authority. Haila's (1999) work has suggested that Shanghai might be moving toward a property bubble. The government's role in regulating and controlling property development and land supply require a serious reexamination.

The situation in Hong Kong also requires a great deal of attention. Statistics suggest that the vacancy rate and number of private dwellings rose quickly in recent years. The number of newly completed private dwellings sold to purchasers (indicated as take-up in Table 3.7) was also lower than the number of completion for the period 1997 to 2001. Amid pessimism on the future of the economy, the high vacancy is likely to continue in 2002 and 2003.

In such a highly capitalized society, property development decisions are all based on feasibility assessment of real estate development companies. Land supply

Table 3.6

Vacant Dwellings in Shanghai, 1994–2000

Year	Vacant dwellings (million square meters)	All types of vacant commodified housing (dwellings, offices, and buildings for commercial use) (million square meters)	Vacant dwellings as a proportion to all types of vacant commodified housing (%)
1994	1.5760	1.7533	89.9
1995	3.4558	4.1339	83.6
1996	5.0726	6.2990	80.5
1997	7.2910	9.6880	75.3
1998	9.0162	12.3833	72.8
1999	9.2223	12.9740	71.1
2000	8.4337	12.5030	67.5

Source: Shanghai Academy of Social Science Housing and Real Estate Research Institute, *A Research Report on Position and Prospect of Shanghai Housing and Real Estate Market*, Shanghai, 2001. www.realestate.gov.cn/manage/uploadfile/market/0000/0521/2001070401.htm.

is always responsive to market signals. A market-led Application List system to supplement the regular land auction and tender program has been put in place in recent years. Hopefully the oversupply of private housing will be a transient phenomenon, and, as the economy picks up again, there will be chances for the vacancies to be cleared. However, the more immediate challenge of the Hong Kong SAR government is to find ways to improve the economy and to create more jobs. The strong state-managed housing system in Hong Kong has witnessed the quickened pace of suspending the sales of HOS flats and replacing it with more interest-free home loans in order to boost up demand for private housing.

In Shanghai, property prices also dropped in 1997 and 1998. Between the third quarter of 1998 and the first quarter of 1999, there were signals showing some bounce back, and from April 1999 onward, prices continued to increase and the rate of increase accelerated in 2000. By late 2001, the average price of commodity housing was at 3,866 yuan per square meter. Thus, between 1997 and September 1998, there was a drop of about 25 percent in property prices in Shanghai.[77] Why are there less discontented property owners? One explanation is that prior to the implementation of the "termination of welfare housing allocation in kind policy," most owners were sheltered from direct loss as the prices they paid for their occupied dwellings were basically heavily subsidized by the work units. One report suggests that over half of the market dwellings were purchased by work units for allocation as welfare housing to employees.[78] In future years, when the Shanghai authority implements the "termination of welfare housing allocation in kind policy,"

Table 3.7

Private Dwellings in Hong Kong—Completions, Take-Up, and Vacancy

	1997	1998	1999	2000	2001	2002	2003
Completion	18,200	22,280	35,320	25,790	26,260	[30,400]	[28,000]
Take-up	15,090	13,050	19,560	29,180	19,320		
Vacancy	35,980	43,820	59,140	54,950	60,410		
%*	3.8	4.5	5.9	5.4	5.7		

Source: Hong Kong SAR Government Rating and Valuation Department (2002), *Hong Kong Property Review 2002*, p. 19.

* Vacancy at the end of the year as a percentage of stock; [ ] Forecast figures

the limited cash subsidy given to employees will pose a new housing affordability problem that will indirectly affect the demand for market dwellings. There is also a need for both governments to reconsider the home ownership promotion policy. Experience from other countries shows that providing affordable housing and decent shelter for all should be the prime concern of government. However, governments should be extremely cautious and not embark on a home ownership policy without taking the right steps to ensure against the risks of home ownership, both for the household and the economy at large.

Notes

1. These are 1999 figures extracted from Hong Kong SAR Government Planning Department's Web site www.info.gov.hk/planning/info_serv/statistic/landu_e.htm.
2. C. Howe, ed., *Shanghai: Revolution and Development in an Asian Metropolis* (Cambridge: Cambridge University Press, 1981).
3. L. W. Pye, Foreword in *Shanghai: Revolution and Development in an Asian Metropolis* (Cambridge: Cambridge University Press, 1981), pp. xi–xvi.
4. L. White, "Shanghai-Suburb Relationship 1949–1966," in *Shanghai,* ed. C. Howe (Cambridge: Cambridge University Press, 1981), pp. 241–68.
5. Estimated home ownership rate provided by Yongyue Zhang, professor of Orient East College, East China Normal University, Shanghai, in a seminar held at City University of Hong Kong on 24 April 2001.
6. Shanghai Residential Housing 1949–1990 Editorial Board, *Shanghai Residential Housing 1949–1990* (Shanghai: Kexue Puji Chubanshe, 1993), pp. 16, 33, 134.
7. Rebecca L. H. Chiu, "Housing" in *Shanghai: Transformation and Modernization under China's Open Policy*, ed. Y. M. Yeung and Y. W. Sung (Hong Kong: Chinese University Press, 1996), pp. 341–74.
8. Tingjun Huang, "More Than 10 Billion Yuan Shanghai HPF Mortgage Loan to Individuals Were Made in 2001," *Shichang Bao* (Market Daily), 14 January 2001.

9. Zhifeng Liu Speech presented at the Tenth Anniversary of the Shanghai Housing Provident Fund Commemorative Meeting, 25 April 2001.

10. Zhengtong Wu, "Housing Provident Fund's Full Implementation for Realisation of Home Owning," Housing for the New Era Conference Papers, Shanghai-Hong Kong Housing Conference, 8–10 January 1997, Shanghai, pp. 54–55.

11. Tingjun Huang, "More than 10 Billion Yuan Shanghai HPF Mortgage Loan to Individuals Were Made in 2001," Shichang Bao (Market Daily), 14 January 2002.

12. Tingjun Huang, "Strong Demand for Loan from Shanghai Housing Provident Fund despite Limited Room for Additional Growth of Accumulated Fund," Jingji Cankao Bao (Reference News on Economy), 22 May 2002.

13. Cheng Cong, "The Strategy of Developing the Housing Provident Fund System—The Case of Shanghai," Chinese and Foreign Real Estate Times, 8 (2001): pp. 12–13.

14. Tingjun Huang, "Strong Demand for Loan from Shanghai Housing Provident Fund despite Limited Room for Additional Growth of Accumulated Fund," Jingji Cankao Bao (Reference News on Economy), 22 May 2002.

15. Yufeng Gao, Summary of speech presented at the commemorative meeting of the tenth anniversary of the Shanghai City Housing Provident Fund, 29 April 2001. See www.realestate.gov.cn/news.asp?recordno=1686&teamno=2&line=111.

16. Zhigang Huang, "Progress and Further Development of Housing System Reform," Housing for the New Era Conference Papers, Shanghai-Hong Kong Housing Conference organized by Shanghai Housing Authority and Hong Kong Housing Authority in Shanghai, 8–10 January 1997, pp. 50–53.

17. Nanfang Fangdichan (Southern Housing and Real Estate), January 2000, p. 30.

18. Shanghai Economy Yearbook 2001, p. 277.

19. Fangdichan Yanjiu yu Dongtai (Housing and Real Estate Studies and Dynamic), 2 (2002): p. 6.

20. Shanghai Housing and Land Resource Bureau, "Building the Housing Consumption System for General Public," see www.cin.gov.cn/fdc/exp/2000092902.htm.

21. Shanghai City Housing System Reform Office, "An Effective Attempt in Marketisation of Public Sale Flats," Zhuzhai yu Fangdichan (Housing and Real Estate), April 1998, pp. 25–27.

22. Anan Shen and Yan Yang, "Meeting the Living Requirements and Improving the Quality of Life," in A Report of Social Development in Shanghai 2001 (Shanghai: Shehui Kehuiyuan Chubanshe, 2001), p. 52.

23. Shanghai Economy Yearbook 2000, p. 352.

24. Fangdichan Yanjiu yu Dongtai (Housing and Real Estate Studies and Dynamics), 2 (2002): p. 5.

25. Zhigang Huang, "Progress and Extension of the Housing System Reform," Housing for the New Era Conference Papers, Shanghai-Hong Kong Housing Conference, Shanghai, 8–10 January 1997, pp. 50–53.

26. See Shanghai Statistical Yearbook 2001, Table 3.8, which gives 11,718 yuan as the average per capita disposal income in 2000. The population census completed in November 2000 shows that the average household size is 2.8 persons. This gives a total annual household disposable income of 32,810 yuan.

27. Monthly rental charge per public dwelling unit is calculated by author: 2.1 yuan x 50 x 0.71 = 74.6 yuan or round-off to 75 yuan. Annual rental charge = 900 yuan.

28. Jing Wang, "An Exploratory Study on the Residential Housing Price to Income Ratio and Rent to Income Ratio in Shanghai," Chinese Housing and Real Estate Studies, 1 (2001): pp. 52–53.

29. See Shanghai Statistical Yearbook 2001, Table 3.9. The per capita spending on ciga-

rettes, wine, tea, and drink in 2000 was 382.01 yuan. A household of 2.8 persons would have spent 1,069.6 yuan per annum.
30. Zhifeng Liu, "Speech by the Deputy Minister of Construction Made at the 2002 National Residential Housing and Real Estate Conference," Shanghai, 4 February 2002.
31. *Shanghai Statistical Yearbook 2001*, Table 3.8.
32. *Fangdichan Yanjiu yu Dongtai* (Housing and Real Estate Studies and Dynamic) 2 (2002): pp. 5–6. Also see *Nanfang Fangdichan* (Southern Housing and Real Estate), 4 (2001): pp. 23–25.
33. Figures are as at March 2002. See *Fangdichan Yanjiu yu Dongtai* (Housing and Real Estate Studies and Dynamic), 3 (2002): p. 5.
34. The Hong Kong Government Planning Department Web site is www.info.gov.hk/planning/index_e.htm. The figures for 1998 were accessed on 5 July 2000.
35. C. Blundell, "Housing: Getting the Priorities Right," in *Managing the New Hong Kong Economy*, ed. D. Mole (Hong Kong: Oxford University Press, 1996), pp. 108–49.
36. E. McLaughlin, "Hong Kong: A Residual Welfare Regime," in *Comparing Welfare States: Britain in International Context*, ed. Allan Cochrane and John Clarke (London: Sage, 1993), pp. 105–40.
37. Consumer Council, *How Competitive Is the Property Market?* (Hong Kong: Consumer Council, 1996).
38. MDR Survey of Housing Aspiration of Households—Final Report (Prepared for the Planning Department of the Hong Kong Special Administrative Region Government) (Hong Kong, 2000), Table 7, p. 23.
39. Kwok Yu Lau, "Housing Privatization: Some Observations on the Disengagement and Delimitation Strategies," in *Housing Express* (Hong Kong: Chartered Institute of Housing, 2001), pp. 11–16.
40. Kwok Yu Lau, "The Implications of the Inadequate Supply and Inequitable Public Sector Housing Allocation Policies," in *Commentary on Housing Policies in Hong Kong*, ed. P.K. Kam et al. (Hong Kong: Joint Publishing Company, 1996), pp. 62–66.
41. Information provided to the author by the Hong Kong Housing Authority on 16 June 2000.
42. Households with monthly earnings between HK$30,001 and HK$60,000 (as at late 1996) are eligible to apply for the low-interest loan or housing under the Sandwich Class Housing Scheme in Hong Kong.
43. Households with monthly earnings below HK$60,000 (as at mid-1999) are eligible to apply for a low-interest loan under the Home Starter Loan Scheme in Hong Kong.
44. Figures on the Home Starter Loan Scheme were obtained from Hong Kong Housing Authority, Paper HOC 47/2002, 2 September 2002.
45. *Hong Kong Housing Authority 1995/96 Annual Report* (Hong Kong Housing Authority, 1996), p. 74.
46. Information provided by the Hong Kong Housing Authority, February and September 2002.
47. Hong Kong Housing Society Web site (www.hkhs.com/) accessed in June 2000.
48. Information provided by the Hong Kong Housing Authority in September 2002.
49. Hong Kong Housing Authority Press Release, 29 August 2002.
50. The income limits had been further relaxed since 1993, and tenants were only required to pay double rent if their household income exceeded three times the prevailing income limits of public rental housing.
51. Kwok-yu Lau, "Safeguarding the Rational Allocation of Public Housing Resources in Hong Kong: A Social Policy Discussion," *Asia Pacific Journal of Social Work*, 7, no. 2 (1997): pp. 97–129.

52. Income includes an element of 5.26 percent of the income limits normally shown in HKHA's publicity materials. This is in recognition of the applicant's monthly contribution to the Mandatory Provident Fund Scheme.

53. Information provided by the senior statistician of the Hong Kong Housing Authority in May 2002.

54. Rebecca L.H. Chiu, "Housing," in *Shanghai: Transformation and Modernization under China's Open Policy*, ed. Y.M. Yeung and Y.W. Sung (Hong Kong: Chinese University Press), pp. 341–74.

55. *Shanghai Economy Yearbook 2000*, p. 352.

56. Jiajin Xie, "Concluding Speech" (presented at the National Conference on Management of Housing Provident Fund and at the National Seminar on Dwellings and Real Estate Information, 30 October 2001b).

57. Shanghai Academy of Social Science Housing and Real Estate Research Institute, *A Research Report on Position and Prospect of Shanghai Housing and Real Estate Market* (Shanghai: Shanghai Academy of Social Science, 2001).

58. See Hong Kong Housing Authority Paper HA34/2002, 21 June 2002.

59. T. Bellman, Housing Reform: The Engine of Growth, PowerPoint Slides, 14 May 2002. Available at www.joneslanglasalle.com.hk/Research/2002/HSBC%20Shanghai%20Conf%20-%20May2002.pdf.

60. Daning Wan, Yongyue Zhang, and Bogeng Chen, "Research on the Contribution Percentage to Shanghai Economic Development Due to Residential Consumption," *Chinese Real Estate Studies*, 1 (2000): pp. 81–96.

61. *Shanghai Statistical Yearbook 2001*, Table 20.56 (Internet edition).

62. Bong Yin Fung, *A Century of Hong Kong Real Estate Development* (Hong Kong: Joint Publishing [HK] Ltd., 2001), p. 297.

63. *Ta Kung Pao*, 6 November 2001.

64. Hong Kong Mortgage Corporation Limited Web site: www.hkmc.com.hk/. The share of residential mortgage loans in the total domestic loan portfolio rose from 9.9 percent to 22.1 percent in the fifteen years to 1995.

65. Calculation done by author based on information provided by the Hong Kong Housing Authority (various years).

66. Speech titled "A Future of Excellence and Prosperity for All" by the Honourable C.H. Tung at the Ceremony to Celebrate the Establishment of the Hong Kong SAR of the PRC on 1 July 1997.

67. Hong Kong SAR Government, *Home for Hong Kong People into the 21st Century—A White Paper on Long Term Housing Strategy in Hong Kong* (1998), paragraph 5.1.

68. See Hong Kong Bank, Hang Seng Bank, and Standard Chartered Bank monthly economic reports published in 2002.

69. Bochu Ye, ed., *Shanghai Housing 1949–1990* (Shanghai: Shanghai General Science Publishing House, 1993), p. 152.

70. The figures for 1993 information were cited in Anan Shen and Yan Yang, "Meeting the Living Requirements and Improving the Quality of Life," in *A Report of Social Development in Shanghai 2001*, ed. Jizuo Yin (Shanghai: Academy of Social Science Publishing House, 2001), pp. 50–51.

71. *Shanghai Statistical Yearbook 2001*, Table 20.56.

72. Shanghai City Housing Development Bureau, "Speed Up Modernization Steps on Housing Industry to Promote Sustainable Housing Development in Shanghai," www.cin.gov.cn/fdc/exp/2001102307.htm.

73. Statistical Reports on Shanghai Municipal National Economic and Social Development 2000 and 2001, http://tjj.sh.gov.cn/tjgb/tjgb.htm and http://tjj.sh.gov.cn/tjgb/gb2001.htm.

74. Hongming Zhang, Daotong Liu, and Meihua Gong, "Studies on Shanghai Housing Demand beyond 2000," *Chinese Real Estate Studies*, 2 (2000): p. 108.
75. Hong Kong Housing Authority, *Housing Authority Performance Indicators*, March 2001 edition and editions of earlier years.
76. Daning Wan, Yongyue Zhang, and Bogeng Chen, "Research on the Contribution Percentage," pp. 81–96.
77. Pei Cui, Outline of Presentation in the seminar on "The Development of Housing Policy, Real Estate Policy and Property Market in Shanghai," Department of Public and Social Administration, City University of Hong Kong, 5 July 2002.
78. Feng Dong, "Current Position and Problems of Housing Allocation System Reform in Some Shanghai State-Owned Enterprises and Institutions," *Housing and Real Estate*, 53 (August 1999): pp. 9–12.

$$4$$

Education

David Chan, Ka-ho Mok, and Tang Anguo

The Impact of Globalization and the Changing Philosophy in Educational Governance

During the past decade or so, scholars have been debating about the impacts of globalization on economic, political, societal, and cultural fronts, and particularly so in the sphere of education (Hirst and Thompson, 1995; Unger, 1997; Currie and Newson, 1998; Jones, 1998; Mason, 1998; Spring, 1998; Dale, 1999; Mok, 2000; Deem, 2001). However, globalization is not a homogeneous process, nor are its effects homogeneous. It has been argued by some scholars that globalization does constitute a new and distinct form of relationship between nation-states and the world's political economy, but that it may take many different forms. In general, the effects of globalization have somewhat eroded both the individual states' capacities to control their own affairs, as well as their mutual arrangements for the collective management of their common interests (Dale, 1999).

One of the major impacts of the process of globalization is related to the fundamental change in the philosophy of governance, and the way that the public sector is being managed. The considerable questioning of the state's ability to continue monopolizing the provision of public services in recent decades has led to the transformation of the states from being "big government, small individual" to the trend of "small government, big individual" (Flynn, 1997). As modern states are very much concerned with better performance in the public sector, fashionable jargons such as "excellence," "increasing competitiveness," "efficiency," "effectiveness," and "accountability" have been introduced, and different strategies such as internal audits, quality assurance, performance pledges, and management-by-objectives have been adopted in trying to improve the efficiency and effectiveness of public services (Mok, 1999).

In spite of the fact that different countries may have chosen various coping strategies in response to this global trend, it is quite evident that there are several ways that globalization will have its impacts upon the public administration and management of individual countries. Politically speaking, the different states' reactions to the changing global circumstances may take two broad forms: *individually*, they may have taken on the form of a "competitive state" (Cerny, 1996), while *collectively*, they may have taken the route of setting up a framework of international organizations through which they try to establish their "governance without government" (Rosenau, 1992). As one scholar points out, "[T]hrough the institutionalization of the global economy; through imposition by the international organizations; by increasing interconnection, both formally and informally; by changing the values of both bureaucrats and policy makers; selection of management practices is shaped increasingly by globalization; trans-nationalization of the nation as apparatus" (Baltodano, 1997, pp. 623–26).

In addition, globalization has also altered the conventional relationship between the state and governance. Unlike the practices in the old days, the new vision of governance today conceives modern states as "facilitators" instead of "service providers," which is particularly true when the "welfare state" is turned into a "competitive state." Jones suggests that globalization promotes a distinct "New World Order" where "much of the globalization process came to be dependent on the adoption of reduced roles for government, not only as a regulator but also a provider of public services" (Jones, 1998, p. 1). This "New World Order" is characterized by governments that revamp the role of a government with a cutback in the scope of their work. At the same time, the notion of "public goods" is replaced by the rhetoric of "economic rationalism" whereby customers' choice and the three "Es" (economy, efficiency, and effectiveness) are emphasized (Welch, 1996). Yet, there are limitations in terms of its convergence effects by the globalization process (Unger, 1997).

Wider Policy Background for Educational Reforms in Hong Kong

Historical Development

At the very beginning, after the Second World War, the provision of public education was not really an important public policy that the British colonial government in Hong Kong had to reckon with. Quite a few of the educational services at that time were provided by the charitable non-state sectors of the society. Sheer quantity of education had to be provided for, through private and subsidized schools, due to the huge influx of migrant population from the Mainland. The interesting remark that "providing education not only involves the quantitative business of putting school roofs over students' heads, but also questions of priority for the various levels and types of education as well as questions of jurisdiction. Whose responsibility was it to provide schooling?" (Sweeting, 1993, p. 65) points to the fact that, during the early periods of colonialism, schooling was not being considered by the colonial government to be of paramount importance to Hong Kong.

It was only in 1954 that a Seven-Year Plan on education was considered by the Education Department, which marked the first time the colonial government started to seriously consider education a policy having wider social implications. Again, quoting from Sweeting, "the seven-year plan may be regarded as a fascinating amalgam of the administrative and the practical sides of the Hong Kong government, relying heavily on the support of an equally intriguing alliance of idealists and pragmatists outside the government. It offered solutions to the problems of quantity, but possibly at the expense of exacerbating the problems of quality" clearly delineates the problem of education at that time (Sweeting, 1993, pp. 108–16).

For the next three decades or so, with the further expansion of its population, together with the demographic changes that had occurred, the Hong Kong government had made tremendous progress in terms of its educational funding and thus further expanded its provisions of education from the primary, through the secondary, to the tertiary sectors. Overall, during this period of time, from the mid-1950s to the early 1990s, the quantity of educational provision was still considered the major focus of its attention. By the early 1990s the government started to expand very rapidly on its tertiary sector of education, after the June Fourth Incident that had occurred in the Mainland in 1989, which brought about the problem of the "brain drain" in Hong Kong.

At the same time, concern about quality in public services also reached Hong Kong and was expressed in such initiatives as Governor Patten's Performance Pledges (Lee and Cheung, 1995). This initiative has begun to make a powerful impact on education at all levels in the Hong Kong Special Administrative Region (SAR) since July 1, 1997. As a matter of fact, in Hong Kong, as in many other countries, a concern for quality also has its roots in managerialism and the new public management.

With the return of its sovereignty to China, the Hong Kong SAR Government has been initiating a series of education reforms at various levels. In September 1997, the SAR Government published Education Commission Report No. 7, titled *Quality School Education*, which set the scene for the need of a quality school education, including the inculcation of a quality culture in the school system; the introduction of "school-based management" in trying to achieve the aims of education in an efficient, cost-effective, and accountable manner; the introduction of a mechanism for quality assurance within the school system as well as a framework for raising the professional standards of principals and teachers, together with the possible establishment of a General Teaching Council, a professional body for teachers (EC, 1997).

Policy of Decentralization in Schools

In an attempt to promote "quality education" in schools, the Hong Kong SAR Government has introduced management strategies to the educational sector. Learning from the U.K. experience, Hong Kong introduced the idea of a school-based management model in order to bring about the policy of decentralization (Brown, 1990) in 1991. It is assumed that when a market mechanism is in place in the education system, schools will become responsive and accountable to the public, and thus will opt for better quality performance. Hence to the managerialists, the right choice is to devolve responsibility to the schools. School-based management is a kind of reform to revitalize schools in terms of flexibility, accountability, and productivity in assuring the quality of education.

This is called the School Management Initiative (SMI) by the Hong Kong administrators (EMB, 1991) so as to set the framework for quality in school education. This involves the move of the locus of control for the implementation of programs and policies "down the line" from the center (the Education Department) to the periphery (individual schools). The basic aims of the SMI have been:

1. To clarify the roles and responsibilities at both the school and system levels
2. To enhance principals' and teachers' sense of professionalism and thorough participation in decision making
3. To ensure accountability through self-managing effort to improve learning and teaching
4. To assure quality in education (Pang, 1997, 1998)

In general, it can be seen that quality is to be promoted by the adoption of a series of management-type strategies, such as the creation of development plans, the setting of mission statements and clear objectives, the initiation of appraisal systems for staff, the provisions for rewards and punishments, and their professional development. In a nutshell, the SMI expects that with increased "transparency" of school operations, broadened participation from parents and the community in school management, increased "accountability" of schools to the general public, and the sharing of experience among schools with similar backgrounds or within the same "quality circle," schools would then improve and be inspired by others to continue to strive for "excellence."

This strategy of school-based management is based on a particular model of organization or administration that assumes that decision making is rational and can be carried out in an orderly manner through decentralization. Yet the reality, in most cases, is that school management is subject to the political realities and constraints that are present in each school and in its relations with the school sponsoring body and the Education Department. Indeed, there are serious problems with the managerial approach to education quality when dealing with the human and political dimensions of organizational settings (Tyler, 1985; Tam and Cheng, 1997).

Furthermore, when one talks about accountability, the issue is about who is accountable to whom in relation to what matters. In short, it is about issues of power, control, and authority. So, the crucial issue is whether the government would put forward an "educative accountability" characterized by democratic governance, educative leadership, and communitarian parenting and professionalism, rather than forms of market, political, and productivity mechanisms (MacPherson, 1996). It is, indeed, questionable whether the new form of accountability is toward the professional model or the consumerist model in a quasi-market situation.

In order to promote quality education in Hong Kong across different levels, the SAR Government has introduced management reforms and adopted a "market-oriented" approach in running education. Thus, in his First Policy Address to the Provisional Legislative Council titled *Building Hong Kong for a New Era* on 8 October 1997, the chief executive of the HKSAR Government, Tung Chee-hwa, specially termed the section on education as "uncovering Hong Kong's treasure," which meant that the SAR Government particularly emphasized the importance of education in order to maintain its competitiveness in the global market. In his policy speech, the chief executive had specially allocated HK$5 billion to set up a Quality Education Development Fund so as to encourage innovation, competition, and self-motivated reform in primary and secondary schools. About three-quarters of the fund have already been allocated to sponsor various projects of different natures.

At the same time, the chief executive promised to set up the General Teaching Council within the next two years, to advance the date for graduate posts making up 35 percent of all primary posts from the year 2007 to 2010, and to raise the percentage of students in whole-day primary schools from the previous target of 40 percent to 60 percent by the 2002 school year. Furthermore, he announced that proficiency in written English and Chinese, together with proficiency in spoken English, Cantonese, and *Putonghua* would be the aim for all secondary school graduates. *Putonghua* would become part of the curriculum starting from the new school year of 1998, together with Information Technology (IT). Finally, he asked the Education Commission to begin a thorough review of the whole education system in Hong Kong, from preprimary to tertiary level (HKG, 1997).

Timetable for the Overall Review of the Education System: 1998–2000

With this, the process on the review of the whole education system in Hong Kong started in 1998. First of all, consultative documents on two sets of performance indicators for both primary and secondary schools were circulated to all schools for their feedback and suggestions concerning the assurance of quality education in schools (ED, 1998a, 1998b). In June, the Education and Manpower Bureau put out a consultative document concerning a five-year strategy (from 1998–2003) for the use of IT in education, hoping to provide an average of forty sets of desktop computers for all primary schools and eighty-two sets of desktop computers in all secondary schools by the school year of 1999–2000 (EMB, 1998a), with an average of about two contact hours per week for each primary student and about six to eight contact hours per week for each secondary student. In 2003, these strategies were accomplished.

In July of the same year, another consultative document on the review of the Education Department itself, in order to streamline its organization and management, was also published by the Education and Manpower Bureau and was released to the general public for further consultation (EMB, 1998b). A new scheme of Quality Assurance Inspection (QAI), as part of the new quality assurance mechanism from the Education Department, was launched when the new school year began in September. Then, by October, the chief executive, in his second Policy Address, mentioned that he would allocate HK$20 million to set up a General Teaching Council, and had earmarked HK$630 million of funding for the use of IT in education (HKG, 1998). With this, the Preparatory Committee on the Establishment of General Teaching Council under the Education Commission published its consultative document on the same topic in November (EC, 1998), while the five-year strategy was being confirmed for the implementation of using IT in education (EMB, 1998c).

The *first major review* of the whole education system sparked public debate in January 1999 with the publication of the *Education Blueprint for the 21st Century, Review of Academic System: Aims of Education* by the Education Commission, which basically set out the overall aims of education and the specific aims of early childhood education, of basic school education and of tertiary education, bearing in mind the need to cater to all of the unique developments of individual students, as well as to foster an outward-looking attitude while remaining firmly rooted in the Hong Kong context. The document, for the first time, also outlines the three stages for this whole review exercise: the first stage on establishing the aims of education, the second stage on the review of the academic system, and the third stage on finalizing the recommendations (EC, 1999a).

By September 1999, another consultative document on *Review of Education System, Framework for Education Reform: Learning for Life* was released by the Education Commission as the *second stage of review*, which is mainly on the review of the academic structure, the curricula, and the assessment mechanisms (EC, 1999b). In October, the chief executive again mentioned education as an important social asset in his third policy address (HKG, 1999) and a consultative document on the holistic review of the school curriculum was released by the Curriculum Development Council (CDC, 1999).

It was in February 2000 that a revised version of the performance indicators for both primary and secondary schools was finally in place for the establishment of an internal quality assurance mechanism within the school system (ED, 2000a, 2000b). At the same time, another consultative document on school-based management was also published by the Advisory Committee on School-Based Management under the Education Department (ED, 2000c). By May 2000, a final consultative document titled *Review of Education System Reform Proposals Con-*

sultation Document was published by the Education Commission as the *third and final stage of review* for the overall review of the education system, in order to receive the final stage of public consultation by the end of July 2000 (EC, 2000a).

Final Outcomes of the Educational Reform Proposals

By September 2000, the EC published its final report by outlining the reform strategies for improving the existing education system. Choosing a theme of "Learning for Life, Learning through Life," the EC cited learning as the key to one's future and education as the gateway to Hong Kong's future. Strongly believing that students are the focal point of the entire education reform, the EC stressed the importance of "all-around development" and "lifelong learning."

It is noteworthy here that the call for "quality education" and for the construction of a "learning society" is not only heard in Hong Kong, but is also voiced in places such as Taiwan, Singapore, and Shanghai. There is no doubt that since the 1960s, Hong Kong, Taiwan, and Singapore have had prosperous economic growths. The rapid growth in these places has led the World Bank to regard them as High Performing Asian Economies, and the terms "Asian Tigers" or "Dragons" were adopted to describe them. But what is common to these dragons is the importance of "human capital." In order to maintain their competitiveness in both the regional and global markets, the governments of these places have attached great significance to the role of education. Thus, the proposal to create a "Learning Society" in Hong Kong is closely related to its intention to strengthen the "global competence and competitiveness" of its citizens (EC, 2000b).

The final outcome of the overall review on the education system was announced by the Education Commission on 28 September 2000 (EC, 2000b). The overall aim of providing education in Hong Kong SAR for the twenty-first century is to build an education system that is conducive to lifelong learning and all-around development of the individuals, with the priority to enable students to enjoy learning, enhance their effectiveness in communication, and develop their creativity and sense of commitment. The vision of the Education Reform is to (1) build a lifelong learning society, (2) raise the overall quality of students, (3) construct a diverse school system, (4) create an inspiring learning environment, (5) acknowledge the importance of moral education, and (6) develop an education system that is rich in tradition but cosmopolitan and culturally diverse.

Although the principles of the reform are student-focused—"no-losers" in the system, quality education, life-wide learning, and society-wide mobilization—its main objectives are to reform the admission systems and public examinations so as to break down barriers and create room for all, reform the curricula and improve teaching methods, improve the assessment mechanism to supplement learning and teaching, provide more diverse opportunities for lifelong learning at senior secondary level and beyond, formulate an effective resource strategy, enhance the professionalism of teachers, and implement measures to support frontline educators.

In terms of the various levels of education, the following are the main areas of concern for the Reform movement:

1. *Early Childhood Education.* It was recommended to establish a single authority to monitor all kindergartens and child-care centers and to raise the qualifications of their teachers.
2. *Basic Education.* In order to lessen the examination pressures for students, it was recommended that (a) the written Primary One examination and interview will be banned; (b) the discretionary places for Primary One will be cut from 65 percent to 20 percent in five years' time, and up to 10 percent should be reserved for applicants outside of the school district; (c) the Primary Six Academic Aptitude Test will be replaced by assessment of school academic performance; (d) basic competency assessments in Chinese, English, and mathematics will be introduced at various stages of basic education; (e) the bandings of five categories will be reduced to three categories for the junior secondary level; (f) nine years of basic education will remain as is; and (g) a system of "through-train" for primary and secondary schools will be introduced whenever possible.
3. *Senior Secondary Education.* It was recommended that (a) a three-year senior secondary schooling be proposed, and a working group will submit its final recommendations to the government by 2002; (b) a broad-based curriculum should be provided in order to avoid premature streaming; and (c) the fine grades in the Hong Kong Examination of Education (HKCEE) and Hong Kong Advanced Level Examination (HKALE) will be abolished by 2002 to be replaced by a new public examination on Chinese and English.
4. *Tertiary Education.* It was recommended that universities (a) should adjust their length of study for certain first-degree programs; (b) have to resolve their own resource-related problems; (c) should overhaul their existing admission mechanism to give due recognition and consideration to students' all-around performance; (d) should install a transferable and articulated credit-unit system among institutions and departments so as to allow room for more flexibility in university education; (e) should increase the numbers of research postgraduate places and self-financed postgraduate places; and (f) society should provide a conducive environment for the development of private university/ higher education institutions in the territory.
5. *Continuing Education.* It was recommended that (a) multiple channels of continuing education, such as distance-learning, community colleges, and so on, should be established; (b) society-wide lifelong learning centers should be set up; (c) a database for continuing education should be estab-

lished; and (d) a working group should be set up to advise on policies of continuing education.

It was in his Fourth Policy Address titled "Serving the Community, Sharing Common Goals" on 11 October 2000, that the chief executive of the HKSAR again emphasized education reform in a section titled "Holistic Education for the New Century," in which he outlined his vision and commitment to education (HKG, 2000, pp. 17–27), including some of the following major items:

1. Increase of over 2,800 teaching posts, and thus improve the teacher-pupil ratio by 10 percent to 1 : 21.8 in primary schools and 1 : 18.7 in secondary schools.
2. Percentage of whole-day primary school places has been increased from 20 percent to 40 percent; hope to achieve the ultimate goal of providing whole-day schooling for all primary students by year 2007.
3. Plan to provide about 6,000 more subsidized school places by 2002–2003 school year so that all Secondary 3 (Grade 9) students of public sector schools can be provided with subsidized Secondary 4 (Grade 10) places or vocational training (at present, only 85 percent can do so) which will incur an extra HK$740 million in recurrent expenditure by 2007–2008.
4. At the senior secondary school level, the intention is to increase the range of learning options to allow students more choice of schools and subjects (such as the use of either Chinese or English as the medium of instruction, the recategorization of schools into three bands instead of five, the introduction of the "direct subsidy scheme" for schools, etc.) according to their own abilities and interests. However, during the processes of implementing the policy, other challenges and difficulties pop in and thus new problems occurred.
5. HK$10 billion has been earmarked for the completion of the School Improvement Program within the next two years.
6. The objective in tertiary education is to have 60 percent of senior secondary school leavers receive tertiary education within the next ten years (at present, only about 30 percent can do so, with 18 percent attending universities, and another 12 percent in other post-secondary/vocational institutions, while a similar amount of students have gone overseas for their further education), thus providing about 28,000 additional places for tertiary education, bringing the total number to around 55,000.
7. A diverse, multilevel, multichannel system of tertiary education, with a common transferable credit-unit system across the board, will become accessible to all, so that lifelong learning will become the norm of society in the future.
8. Full implementation of all the plans will increase the annual recurrent expenditure on education by about HK$2 billion.

The challenge that Hong Kong faced after the Asian financial crisis was to have a balanced budget, but, most unfortunately, it had huge deficits for the financial years of 2000 and 2001. Hence, some of these grand ideas have been put to a halt. At the same time, when policies are implemented, there will be many hurdles to go through, especially from the frontline workers (in this case, the teachers themselves) in order to achieve the expected outcomes. This proved particularly true when the language policy and assessment schemes were not well received by the teachers and parents alike. With these main backgrounds on the Hong Kong scene, we will now try to have a look at education reform in Shanghai.

Changing Educational Policy in Shanghai, China's Dragon Head

After the founding of the People's Republic of China, the communist state adopted a centralization policy in the education sector. Education was regarded as an instrument to spread the official ideology of communism; thus the communist state exerted strict control over education. As a result, an educational system was developed that was characterized by a unified system of planning and administration, curriculum structure, syllabus and textbooks, student enrollment, and allocation of university places and employment in the 1950s. It was in this way that all schools and tertiary educational institutions were under the direct and strict control of the party-state. The state assumed the responsibilities to formulate educational policies, allocate educational resources, exert administrative controls, recruit teaching staff, and decide on curricula and textbooks. In a nutshell, the party-state monopolized the provision, financing, and governance of education services.

However, during the period of the Cultural Revolution, from 1966 to 1976, there were many disruptions in both basic education and tertiary education, with many students and young people being sent to the hills and villages for labor. It was only after 1978, when Deng Xiaoping was restored to power, that the Economic Reforms and Open Door Policy were established such that education was once again being considered as the most important institution to bring the nation back to its right track so as to further boost its economy. What is sociologically significant, in the case of China, is the fact that there was a shift from a centrally planned economy to a "socialist market economy" which has somehow redirected China's education toward a "market-oriented" approach.

The promulgation of the "Decision of the Central Committee of the Chinese Communist Party of China on the Reform of the Educational System" in 1985 marked the first critical step to decentralize Chinese education. Admitting the too-rigid governmental control of schools and its inefficient management, the Decision set out the general guidelines for decentralizing China's educational system. Under the principle of linking education to economic reform (Zhou, 1995), the document called for devolution of power to lower levels and reducing the rigid governmental controls over schools (CCPCC, 1985).

In response to the 1985 "Decision on Reform of the Educational Structure," Shanghai has emphasized educational development in its research strategies ever since June 1986. Educational development was placed as a key research item in the research plan for "philosophy and social sciences" during the period of the city's "Seventh Five-Year Plan" (1986–1990). The research has further shown that Shanghai's strategic goal for educational development is having its own local-specific orientation: the "economies of scale" for Shanghai's educational development must be appropriate for its own economic and scientific capacities, as well as for its own social and political cultures.

Furthermore, in order to spearhead the national goal of the "Four Modernizations," it is realized that the quality of education in Shanghai must reach and maintain a "first-class standard" within the country and that certain schools or disciplines must meet the "internationally recognized" advanced standards. In order to establish an open, pluralistic, and modernized educational system within a socialist society with a reasonable structure, sound functions, and a flexible, well-connected, and well-managed system, Shanghai will need to become a "national city" with an advanced level of educational development, as well as an open base for educational exchanges with foreign countries (Task Force on the Strategy of Education Development in Shanghai, 1989, p. 16).

To reassure the reform policy on the decentralization of education, the CCP and the State Council further issued the "Program for Education Reform and Development in China" in 1993. This document declared that "the national policy is to actively encourage and fully support social institutions and citizens to establish schools according to laws and to provide right guidelines and strengthen administration" (CCPCC, 1993). In this way, the Central Government had changed its policy from direct control to managing schools through legislation, funding, planning, advice on policies, and other necessary means (SEC, 1993, p. 6). With the retreat of the state in the educational sector, particularly in terms of the reduction of state provision, state subsidy, and regulation (Le Grand and Robinson, 1985), an "internal market," or "quasi-market," is slowly emerging in China (Le Grand and Bartlett, 1993; Mok, 1997; Mok and Chan, 1998, 2001).

In 1993, after the issuance of the "Program for Education Reform and Development in China" by the central government, Shanghai started to conduct research on how to step forward toward the twenty-first century, in order to chart its new course with a new round of strategies in its economic and social developments. In terms of education, according to the research, its hope for development is that "with the request of 'three centers,' Shanghai will build up a 'first-class' lifelong educational system in Asia, with reasonable structure in promoting an all-round personal development, so as to improve the quality of its citizens, and to cultivate a large pool of talents who are able to meet the needs of its economic and social developments in the first half of the 21st century, and to satisfy the increasing needs of education for its citizens" (Leading Group of the Project on Shanghai into the 21st Century, 1995, p. 447). One of the specific goals is to form a great modernized educational system that links up the various sectors of the preprimary, primary, secondary, and tertiary levels, which can also communicate with the various aspects of general education, vocational, and adult education (Cai, 1995, p. 353).

At the same time, in order to have more diversity of educational provisions, the non-state sectors are strongly encouraged by the various levels of government to deliver educational provisions by setting up *minban* (people-run) schools and tertiary institutions as providers, with the emergence of self-financing students as the customers in the education marketplace (Mok and Chan, 1996; Chan, 2001; Chao, 2001; Chan and Mok, 2001a, 2001b). It is under such wider socioeconomic and policy contexts that China's education has been undergoing the process of decentralization through the strategy of marketization (Mok, 2000). Marketization in the Chinese context is meant to be a "process whereby education becomes a commodity provided by competitive suppliers, educational services are priced and access to them depends on consumer calculations and ability to pay" (Yin and White, 1994, p. 217).

Being affected by strong market forces and the policy of decentralization, China's education has made fundamental changes in terms of orientation, financing, curriculum, and management (Agelasto and Adamson, 1998). Indeed, these changes that are taking place in China's educational sector are in tune with the overall trend in China's social welfare sector since the reform era, which has reemphasized individual responsibilities and local initiatives, thus lessening the importance of the state in welfare provisions (Wong and Flynn, 2001).

Shanghai's Educational Reform Strategies

By the mid-1990s, Shanghai further proposed the idea of constructing itself as a "first-class city with first-class education" (Su and Chen, 1994, p. 6) in order to achieve the economic goal of establishing Shanghai as an international economic, financial, and trade center ("three centers"). In so doing, Shanghai has defined its educational goal as "a first-class lifelong educational system in Asia with reasonable structure and benefits to an all-round personal development" (Leading Group of the Project on Shanghai into the 21st Century, 1995, p. 447).

In February 1996, the then Mayor of Shanghai, Xu Kuangdi, proposed six measures in response to the "security of investment and conditions for education," which include, among others, the building up of a supervisory system for the implementation of educational expenditures and the levying of additional surcharges for social enterprises (Editorial Board, China Education Yearbook, 1997, p. 456). In 1997, the Shanghai government formulated its ambitious plan in developing a system of first-class elementary education in Shanghai. The plan was titled the "Establishment of First-class Elementary Education and Year 2010 Prospect," which was formulated and implemented under Shanghai's "Ninth Five-Year Plan" (Shanghai Municipal Commission of Education, 1998, p. 139). While the study on the strategic planning of educational development can highlight the future direction of education in Shanghai, the investment in education will be the material base for the realization of its major priority in Shanghai's educational development.

In 1999, the Shanghai Municipal Government decided to further increase its educational investments. During the four years from 1999 to 2002, the Municipal Government decided to have an average increase of "2 percent per year" on its educational expenditures at the municipal level. Concurrently, the total investment for educational expenditures was 4 percent of Shanghai's total GDP (Shanghai Municipal Government, 1999, p. 129). By September of the same year, following the National Conference on Education Work held in June in Beijing, a Shanghai Meeting on Education Work was also convened by the Shanghai municipal authorities. Based upon the experiences of Shanghai's economic reform and educational development in the 1990s, the meeting further discussed the goals and main tasks of educational development in Shanghai by the early twenty-first century.

In the 1990s, the social forces in Shanghai started to provide education in the formal sector of educational institutions. The first *minban* school after the Cultural Revolution was established in July 1992. In August of the same year, the first *minban* university called the Sanda University was also formally established (Chan and Ngok, 2001, p. 68). While utilizing "social forces" to establish schools, Shanghai also cooperated with overseas organizations for its school establishments, such as international schools (Cai, 1995; Su and Chen, 1994).

From 1993 onward, Shanghai piloted a scheme to transform the management system of the public primary and secondary schools (*zhuanzhi xuexiao*). Under this scheme, with the ownership of the public schools still in the hands of the government, their administration may be contracted out by the administrative departments of education to enterprises, business organizations, social organizations, or individual citizens. The contracted out public schools may be run with reference to the policies applicable to the *minban* schools in respect to student recruitment, collection of tuition fees, selection and appointment of principals and teachers, and schools' internal management. As a result, these schools can now have a relatively higher degree of autonomy in running their own affairs. In general, these kinds of schools can be categorized as "public schools run by social forces."

Currently, Shanghai uses a wide variety of social forces in running formal education, which include some of the following main practices: (1) *minban* primary, secondary and postsecondary schools and colleges established by individual citizens, social organizations, enterprises, and institutions; (2) "public schools run by social forces"; (3) the involvement of various social forces in providing education through community education; (4) various kinds of joint school establishments sponsored by enterprises, or institutions, and primary or secondary schools; and (5) making use of foreign resources and donations from overseas Chinese in providing educational services, as well as cooperation with foreign organizations and individuals in the running of schools, including international schools (Editorial Board, China Education Yearbook, 1997, p. 459).

In Shanghai, by the end of the year 2000, there were altogether 125 *minban* primary and secondary schools, 67 public schools run by social forces (*guoyou minban*), 4 *minban* higher educational institutions endorsed by the Education Ministry, and 12 *minban* postsecondary institutions endorsed by the municipal government, together with more than 300 *minban* kindergartens (Kang, 2001, p. 282). With the involvement of social forces in educational provisions, a new kind of educational investment system has taken shape in Shanghai.

This is a diversified funding structure for education under which the state, society, and individuals all jointly contribute to the financing of education. In Shanghai's current educational expenditure, about 50 percent comes from the government, with another 30 percent from society and communities, and the remaining 20 percent from individuals (Kang, 2001, p. 284). This kind of educational investment system, when contrasted with that of the mid-1980s, reflects the important aspect of fiscal decentralization of education in Shanghai, which in a way can be considered as more "advanced" in terms of the marketization process in the provisions of educational services by the non-state/private sectors as compared with that of Hong Kong.

By January 2001, the Shanghai authorities declared their main goals for educational reform and development for the early part of the new century, as follows: to form an open lifelong educational system with reasonable structure, to construct a modernized educational system that is world-oriented and compatible with the world-class status of Shanghai, and to build Shanghai into a "Learning City" (Shanghai Municipal Commission of Education, 2000; Ngok and Chan, 2000a, 2000b). These new developments can be seen as some of the local responses of Shanghai to the impact of globalization.

Following are some of the more salient features of its educational reform strategies:

Territorial Decentralization

In the context of Mainland China, territorial decentralization involves a redistribution of control among the different geographic tiers of government, such as nation, provinces, county/township, and schools. A transfer of power from higher to lower levels is considered territorial decentralization (Bray, 1999, pp. 208–9). As an intermediate between the central government and the grassroots level, the Shanghai municipal government takes up the responsibility devolved by the central government on the one hand, and makes decentralization policy for the submunicipal levels on the other.

In line with the decentralization of power to the provincial level, the Shanghai Municipal Government and the Municipal Commission of Education began to devolve the power of administrating elementary education among the different geographic tiers of government within its jurisdiction, such as districts, counties, townships, and schools. A series of local administrative regulations has been formulated. The duties and powers of submunicipal governments in developing elementary education are clarified and specified. The districts and counties are encouraged to actively explore, according to law, the new systems for overall development of various forms of local education and the implementation of quality education.

The municipal government has gradually shifted its function from micromanagement to macrocontrol of school governance, while the status of schools as legal entities has been established. Measures are taken to encourage the schools in establishing school-based management: to set their own targets for development, to form their own styles and features of schooling, as well as their own school cultures, so as to stimulate the schools to continuously improve their standards of educational provision, under the prerequisite of carrying out the national educational guiding principles while sticking to the rules and regulations as promulgated by the state for school management.

With the territorial decentralization, the role of governments at the county and district levels in educational financing has also been enlarged. During the whole of the 1990s, governments at all levels in Shanghai invested RMB 468 million in education. Among them, quite a large portion came from the submunicipal levels of government. Some 35 percent, or even 40 percent, of the budget for the district or county governments was invested in education (Kang, 2001, p. 283).

Functional Centralization

Functional centralization refers to a shift of powers between different authorities that operate in parallel (Bray, 1999, pp. 208–9). For quite a while, the administration of education in Shanghai was rather fragmented. There was no single educational administrative unit responsible for all aspects of the public system of education. It used to be that the education administration was shared by three government departments, namely, the Municipal Bureau of Education, which was responsible for general education; the Municipal Bureau of Higher Education, which was in charge of higher education; and the Office of Education and Health under the municipal government, which was responsible for the coordination of education and health affairs. Under such a fragmented administrative system, overlapping of functions, overstaffing, and low efficiency were thus unavoidable.

In order to maintain a more efficient and effective educational governance system, the Shanghai Municipal Government sought to establish a unified educational administrative unit. After several years of major efforts, Shanghai was finally restructured in its education governance with the abolition of the above-mentioned three separate departments. In its place, a single and comprehensive organ, the Municipal Commission of Education, was formed in 1995 to be in charge of all aspects of education in Shanghai. This process can be seen as a means of "functional centralization" or "re-centralization" (Hawkins, 2000), whereby a more efficient and effective school governance system can be put in place for the marketization effort, as well as to curb the losing of too much control from the center to the periphery.

This Municipal Commission of Education is under the leadership of both the Ministry of Education and the municipal government. Under the Municipal Commission of Education itself, there are bureaus of education at the county and district levels, which are mainly in charge of precollegiate education. Furthermore, under the guidance and leadership of the Municipal Commission of Education, submunicipal levels of government, local communities, and educational institutions are all coordinating with each other to provide new forms of educational provision.

The Increasing Role of Local Government in Higher Education

The relationships between the government and universities in China are very complicated. Although almost all universities in China are funded and regulated by the state, they are rather diversified in terms of governance. The Ministry of Education is responsible for regulating the conditions for access to higher education, curriculum structure, degree requirements, assessments, and examinations, as well as the appointment and remuneration of academic staff. However, not all universities are run by the Ministry of Education. In fact, many universities and other higher educational institutions are run and administrated by different departments at the central level, whereas others are under the control of local governments.

Shanghai is one of the regions in China where higher education is most prosperous and well developed. In the early 1990s, there were about fifty full-time colleges and universities in Shanghai. Among them, more than half were run and funded by the Ministry of Education and other departments at the central level; others were run and funded by the Shanghai municipal education department and other departments. Such an administrative system of higher education led to the separation of the center and the locality, the segmentation of universities, resources wastage, and functional overlapping, and thus could not achieve the economies of scale.

According to the reform planning determined by the central government, the objective of reform in higher education system is to set up an administrative system that is characterized by a two-tier management, based on the division of labor between the central and provincial governments and coordinated by the provincial government. In this way, with the decentralization policy laid out by the central government, the Shanghai municipal government strengthened the coordinating function for its local universities, such as the Shanghai Jiaotong University. The municipal government began to make adjustments for the universities and colleges by transferring them to the jurisdiction of the Municipal Commission of Education.

At the same time, many universities and colleges that had similar functions were merged and combined together, which had their precedents back in the 1950s. For example, in May 1994, the original Shanghai University was merged with the Shanghai University of Industry, Shanghai Science and Technology University, and Shanghai Higher Vocational College of Science and Technology to form the new Shanghai University, with their own endowments. Similarly, in October 1994, the Shanghai Normal University and Shanghai Technical Normal Colleges were also merged together.

Furthermore, resources were concentrated and directed to a few key universities and key disciplines so as to enhance the investment efficiency. For instance, starting in 1999, Shanghai decided to pull financial resources together to fully develop Fudan University and Shanghai Jiaotong University, the two main universities that were under the Ministry of Education but were located in Shanghai, aiming to build them into "world-class" universities. Meanwhile, the municipal government started to cooperate with the Ministry of Education to run and fund all seven MOE-led universities located in Shanghai. Besides two special colleges, all the higher educational institutions originally run and administrated by the central departments and located in Shanghai have now been taken over by the municipal government.

Comparison between the Educational Experiences in Hong Kong and Shanghai

The educational reform of "marketization" that has been adopted by the Chinese central government in general, and the Shanghai municipal government in particular, is highly instrumental, intending only to improve administrative efficiency and effectiveness, rather than to make a fundamental shift in value orientation (Mok, 1997; Mok and Chan, 1998). In fact, the private sector still plays a very limited and peripheral role, under the dominance of public institutions, in the provisions of educational services (Hayhoe, 1996).

More sociologically interesting is the fact that the public sector in China's higher education is getting more private. Public universities run businesses and enterprises, charge tuition fees, and develop courses for newly emerging work sectors (Mok and Wat, 1998). The growth of private higher educational institutions, coupled with the adoption of market principles and strategies in recovering educational costs, suggests that the Mainland is moving in a similar, though not identical, trajectory to the global process marketization in education (Mok, 1997). The public/private boundary is becoming blurred in this respect, as China's higher education has been affected by the strong tide of the market economy (Zhou and Cheng, 1997).

On the other hand, working in a policy context in which the quest for quality, efficiency, and effectiveness is emphasized, educational practitioners and academics in Hong Kong found themselves caught up in an "iron cage of rationality," as suggested by Weber (Mitzman, 1971). To seek an objective measurement of quality, as well as to reduce wastage and ambiguity, requires measures to ensure quantification to reward the winners and show up the losers. Unlike the case of the Mainland, the Hong Kong SAR Government has committed herself, both ideologically and practically, to the practice of "managerialism." The recent education reforms in Hong Kong have, undoubtedly, shown that it has been influenced by the tidal wave of marketization in enhancing the efficiency, effectiveness, and economy of education.

In comparing the educational developments of the two cities, eight indicators of (1) gross enrollment ratio, (2) average class size in primary and secondary schools, (3) pupil-teacher ratio for primary and secondary schools, (4) percentage of degree holders serving in primary and secondary schools, (5) government expenditures per pupil per year in both primary and secondary schools, (6) ratio of government expenditure on education to GDP, (7) government expenditure on education as a percentage of total government expenditure, and (8) the distribution of education attainment of population aged fifteen and over have been used by the authors to show their differences.

Indicator 1. Comparison of Enrollment Ratio (Primary)

To be comparable, gross enrollment ratio is used in the comparison because net enrollment ratio in Shanghai is not available. This means that average pupils are included in the figures. Thus, although the average gross enrollment ratio is over 99 percent in Shanghai and even over 100 percent in Hong Kong in the past few years, it does not reflect a situation of full enrollment (see Tables 4.1 and 4.2, p. 112).

Indicator 2. Comparison of Average Class Size in Primary and Secondary Schools

In both Hong Kong and Shanghai, although the average class size has not been reduced in the past few years, it is argued that class condition in Hong Kong is better than that in Shanghai. Average class size in Shanghai is over forty students in general. In contrast, the average class size in Hong Kong is below forty students in general. Besides, average class size is below thirty students in Hong Kong Secondary 6–7, which is significantly smaller than that in Shanghai Senior Secondary (see Tables 4.3 through 4.6, pp. 113–114).

Indicator 3. Comparison of Pupil-Teacher Ratio for Primary and Secondary Schools

Broadly speaking, pupil-teacher ratio in Shanghai is lower than that in Hong Kong for both primary and secondary schools. However, pupil-teacher ratio in Hong Kong has continued to drop since 1994/95. On the contrary, in the past few years, there was no significant improvement in this area in Shanghai. As a result, the range of difference becomes smaller (see Tables 4.7 through 4.10, pp. 115–116).

Indicator 4. Comparison of Percentage of Degree Holders
Serving in Primary and Secondary Schools

There were no statistics indicating the percentage of degree holders serving in Shanghai primary schools before 1999. Such statistics started in 1999, and the figure is 0.57 percent. This reflects that the total number of degree teachers in Shanghai primary schools was rather insignificant until 1999. In contrast, there was a significant growth in the percentage of degree holders serving in Hong Kong primary schools, from 11.53 percent in 1994/95 to 37.71 percent in 1999/2000.

In secondary schools, the percentage of degree teachers had grown continuously since 1995 in both Hong Kong and Shanghai. In 1995, 39.46 percent and 71.69 percent of teachers serving in secondary schools were degree holders in Shanghai and Hong Kong, respectively. In 1999, the figures grew to 49.82 percent and 85.17 percent in Shanghai and Hong Kong, respectively (see Tables 4.11 through 4.14, pp. 117–118).

Indicator 5. Comparison of Government Expenditure per Pupil
per Year for Primary and Secondary Schools

In absolute terms, the government expenditure per pupil in Hong Kong is much higher than that in Shanghai. However, it is known that the consumer price in Shanghai is much lower than that of Hong Kong. Hence, the figures should not be compared in absolute terms. Moreover, government expenditure per pupil only refers to the financial budget (*caizheng bokuan*) of government per pupil, but it excludes other education expenditure from various sources that occupy a significant proportion of educational expenditures in Shanghai (see Tables 4.15 through 4.18, pp. 119–120).

Indicator 6. Comparison of Ratio of Government Expenditure
on Education to GDP

There was a continuous growth in the ratio of government expenditure on education to GDP in both Hong Kong and Shanghai. The ratio, which was 1.44 percent in 1995, grew to 1.74 percent in 1999 in Shanghai. In Hong Kong, the ratio grew from 2.8 percent in 1994/95 to 4.1 percent in 2000/01. However, government expenditure is the major source of educational expenditure in Hong Kong. In contrast, apart from the financial budget (*caizheng bokuan*) of the government, other sources occupy a significant proportion of educational expenditure in Shanghai. Hence, these figures cannot reflect a comprehensive picture of the investments on education in the two cities (see Tables 4.19 and 4.20, p. 121).

Indicator 7. Comparison of Government Expenditure on
Education as a Percentage of Total Government Expenditure

In the past few years, government expenditure on education continued to grow in both Shanghai and Hong Kong. However, government expenditure on education as a percentage of total government expenditure in Shanghai kept at a level of 13 percent, whereas the percentage grew from 17.4 percent in 1994/95 to 23.4 in 2000/01 in Hong Kong. While there was a rapid growth of total government expenditure in Shanghai all through the years, there was only a minor growth in Hong Kong in the last two years. Thus, such a comparison reflects that the Hong Kong government had invested more resources on education than the Shanghai government in the past (see Table 4.21 and 4.22, p. 122).

Indicator 8. Comparison of Distribution of Education
Attainment of Population Aged Fifteen and Over

This comparison reflects that the problem of illiteracy and semi-illiteracy in Hong Kong was more serious than in Shanghai. And the percentage of tertiary education was of a similar nature. As the educational systems in the two cities are rather different, we have combined the percentages of junior and senior secondary in Shanghai, so as to compare them with the percentages of secondary/matriculation in Hong Kong. Then we find that over 60 percent of the people in Shanghai had attained secondary education, whereas there was only 45–47 percent of the same cohort in Hong Kong. Accordingly, the general education level of people in Shanghai was higher than that in Hong Kong. Nevertheless, it should be pointed out that the percentage of "college and high level" in Shanghai does not separate between the percentages of "degree" and "nondegree." In this regard, it is quite possible that the percentage of "degree" in Hong Kong might be higher than that in Shanghai (see Tables 4.23 and 4.24, pp. 123–124).

In giving a general picture of the educational developments of both cities, in terms of their educational achievements, expenditures, and so forth, it can help us have a better idea of the similarities and differences between Hong Kong and Shanghai. However, besides the microaspects, we need to probe into the macro-dynamics of their fundamentals in order to have a much broader perspective to review them.

In comparing the experiences of both places, we can argue that Hong Kong's education has definitely been affected by the strong tide of managerialism. Recent education reforms are part and parcel of the public sector reforms initiated in the territory since 1989. More interestingly, the initiation of a "quality assurance" movement and the obvious shift to a far more "management-oriented" approach in Hong Kong's education have clearly demonstrated the popularity of the ideas of "corporate management" in shaping and managing the educational sector in the territory.

In this regard, the recent education reforms in Hong Kong can be understood as a response to the concerns raised in the local community about whether academic quality can be maintained, especially after turning from an elitist system to a mass education. Most important of all, the adoption of a more "management-oriented" approach in running education in Hong Kong must be analyzed against the wider context of "managerialism" in which performance in terms of measurable and quantifiable outputs is emphasized, and downsizing, privatizing, and corporatizing the government are becoming more popular (Clarke, Cochrane, and McLaughlin, 1994; Hughes, 1994).

Although recent reforms in the educational system in Shanghai seem to suggest that China's education has been going through a similar global experience of marketization, a closer scrutiny clearly indicates that the Chinese counterparts are far more "instrumental" when adopting market mechanisms, which are intended as a measure to improve administrative effectiveness, rather than to make a fundamental shift of value orientation toward notions of public choice. Unlike the case of Hong Kong in which marketization of education has shifted to a "corporate management" approach, the Shanghai experience can be argued as the government's attempt to make use of market forces and new initiatives from the non-state sectors to create more learning opportunities, instead of an ideological shift to managerialism and its practices, showing that Shanghai and Hong Kong are, in fact, "One Country, Two Systems."

Thus, in comparing and contrasting the marketization projects in both Hong Kong and Shanghai, we must differentiate the Shanghai experience of marketization from that of Hong Kong. Clearly commercial aspects, user pays, and a limited role for private provision in Shanghai suggest the reduced state role in educational provision and financing, but this process does not constitute a withdrawal of state control. Despite the fact that the state has tried to "roll back" as a direct provider, it tries to facilitate and revitalize other non-state sectors, and even the market itself, in terms of educational provisions, as the state realizes that the government alone could never meet people's pressing educational needs (see Table 4.25, pp. 125–126).

A Global Trend on the Marketization and Privatization of the Educational Sector

In recent decades, there has been considerable questioning of the state's ability to manage a monopoly of public services. In the West, some people have been arguing that the welfare state is in crisis (Habermas, 1976; Le Grand and Robinson, 1985). Realizing the importance of productivity, performance, and control, governments have begun to engage themselves in transforming the way that services are managed (Flynn, 1997). The traditional "public administration" is now giving way to the "new public management," the paradigm shift which marks a move toward a more transparent and accountable public sector. There is a fundamental change in the philosophy of governance, shifting from the traditional paradigm of "public administration" emphasizing the monopolistic role of the state in social provisions to the "management-oriented" approach that centers around the notion of "economic rationalism," which is the belief that economic factors are the most important ones that shape social and public policies and drive individual actions, embodies a universal model of rationality in shaping social policy in modern days (Rees, 1995; Bates, 1996; Dudley, 1996).

Central to the notion of "new public management" is managerialism and marketization of public service. While the aim of managerialism is to introduce a more effective control of work practices (Pollitt, 1990), that of marketization of public services is to develop market mechanisms and to adopt market criteria within the public sector (Hood, 1991). Giving due importance to the "free market" and "individual choice," the role and function of the state should be reduced because "marketization" is seen as inherently more economically "rational" (Grace, 1994). Adhering to a positivist view in economics and perceiving economics as a science, a new economistic rhetoric of individual rights, efficiency, and choice has become more fashionable in public sector management (Aucoin, 1990; Pusey, 1991). In practice, the notions of doing more with less, productivity gain, and working smarter with fewer staff and less resources (Ball, 1998; Welch, 1998) have unquestionably marginalized the noble goal of the social good, neglecting moral concerns about the quality of life for people in general (Bates, 1996; Rees, 1995).

This ideological commitment to "economic rationalism," that is, "the domination of social policy by the language and logic of economics" (Welch, 1996, p. 4) has exerted a significant impact on education. Thus, by using "market principles" and introducing competition to the educational sector, people believe the delivery of educational services would become more efficient, and thus a higher quality of education could be maintained. In this way, the school authorities in the United Kingdom, the United States, and Australia have tried to give more choices and options to students and their parents, and have thus adopted market principles and introduced "competition" to the educational arena, so as to improve the quality of their educational provisions.

It is noteworthy that the economistic rhetoric of individual rights and ideologies of "efficiency" is gaining momentum, not only in the developed countries, but in the developing countries as well (World Bank, 1988a, 1998b; Bray, 1996a). Levy (1986) reports that in Latin America, private schools are challenging the dominance of public schools. Other scholars have also noted that different countries, regardless of their levels of development, have a mixture of both public and private educational services (James, 1992; Tilak, 1991; Bray, 1997). In fact, many empirical studies have revealed that the share of the private sector in providing educational services is becoming more significant than public and state-run education (World Bank, 1996; Yee and Lam, 1995; Mok, 1997). Bray's work (1996b) on the financing of education in East Asia also discovers that parents and local communities have played more important roles in the financing of education, while the state has gradually withdrawn from the frontier of educational provision. Similar experience is also reported in Mainland China, leading to a new definition of the state/society relationship and public/private boundaries (Mok and Wat, 1998; Cheng, 1995). In short, unlike the traditional notion of education, which was dominated by the public sector, the emergence of "internal markets" and the prominence of "economic rationalism" in the educational sphere seem to be features of the globalization agenda.

As discussed earlier, the impact of globalization, crises of the state, the search for a post-Keynesian settlement, fragmentary impulses in society, and the contradictions of identity politics have created immense pressures for the restructuring of the state. With increasing constraints on public expenditure, the public management today is much more concerned with how to reduce, or at least prevent, cost from continuing to rise. Such concerns have inevitably changed the nature of government from service provider to regulator, thus modern states are conscious about how to regulate the quality of public services. As Le Grand and Bartlett (1993) suggested, modern governments have become basically service purchasers, "with state provision being systematically replaced by a system of independent providers competing with one another in internal or 'quasi-markets'" (p. 125).

In short, the "welfare crisis" that was led by economic globalization has turned governments from a "primarily hierarchical decommodifying agent" into a "primarily market-based commodifying agent" (Cerny, 1996, p. 256). Not surprisingly, the replacement of the welfare state by the "competitive state" has also transformed the state from maximizing its general welfare to maximizing its returns on investment, and thus brought about greater managerialism among the public services. Nowadays, the public sector is not so much about the delivery of public values as about the management of scarce resources.

Under such a global tidal wave, modern states have begun to transform the ways they manage themselves. Notions such as "reinventing government" (Osborne and Gaebler, 1992) and "entrepreneurial government" have become far more fashionable, and the concomitant consequence is the initiation of reforms in the public sector management. In order to improve the efficiency and effectiveness of public service delivery, new ways to maximize productivity and effectiveness comparable to that of the private sector are sought. With emphasis given to effectiveness, efficiency, and economy in the delivery of public services, marketization and privatization are the two major strategies commonly adopted by governments in different countries in the running of public services (Walsh, 1995).

Despite the fact that there are different interpretations of these two "terms," what are common to them is (1) the belief in the ideological commitments of neoliberalism, which holds the view that the state should not bear the primary responsibility to serve all public-good functions (Dale, 1997), and (2) the recognition of the state's severely limited capacity to act in certain policy areas (Offe, 1990). In this respect, "privatization" is closely associated with the reduction in state activities, especially in terms of the reduction in state provision, state subsidy, and regulation (Le Grand and Robinson, 1985). Yet, "some writers describe privatization as another form of decentralization. Certainly privatization may be a form of decentralization in which state authority over schools is reduced. However, it is not necessarily decentralizing. Some forms of privatization concentrate power in the hands of churches or large private corporations. In these cases, privatization may centralize control, albeit in non-governmental bodies" (Bray, 1999, pp. 208–9).

In essence, privatization is concerned with the transfer of responsibility originally shouldered by the state to the non-state sector (Johnson, 1990), or with a change of the nature of the government involvement (Foster, 1992). It is under such a wider socioideological context that the demands of "structural adjustment," that is, the diminution of the public sector and the expansion of the private sector, have become increasingly keen in contemporary society. Turning into concrete policy terms, different ways and measures along the line of marketization and privatization/decentralization have been introduced and implemented with the intention to lessen the financial burden of the state and to improve the performance of the public sector (Flynn, 1997; Hirst and Thompson, 1995; Hood, 1995). Corporate management, devolution of power and authority, the role of the markets, and the popularity of the ideology of neoliberalism are some of the key elements for the restructuring of the state in the public domain (Taylor, Rizri, Lingard, and Henry, 1997; Welch, 1998).

Education, being one of the major public services, is not immune to this global tidal wave of marketization and privatization. Central to all of this educational restructuring is a new concern with the principle of quality in education. Schools, universities, and other learning institutions now encounter far more challenges and are being subjected to an unprecedented level of public scrutiny. The growing concern for "value for money" and "public accountability" has also altered the effect that people's value expectations have on education. All "providers" of education today inhabit a more competitive world where resources are becoming scarcer; yet at the same time, providers have to accommodate increasing demands from the local community as well as the changing expectations of parents and employers—the "customers." In order to be more responsive to all those competing demands, an emphasis on quality assurance has been introduced into schools and universities, within an overarching paradigm of "economic rationalism" in educational services. Within such a policy context, various educational institutions have adopted market principles and introduced "competition" to the educational realm, with the popular belief that education should serve economic purposes (Slaughter and Leslie, 1997).

It is within this policy context that education has become a tool to accomplishing an "integrated world system" along market lines similar to what scholars have suggested on the existence of the educational marketplace (Ball, 1998; Bridges and McLaughlin, 1994). Marketization and privatization in the educational sector are often considered instruments of economic and social policy, as in the case of a "user-pays" principle in education, and in other social services as well. More important, the recent changes and transformations that are taking place in the educational sector in various countries seem to suggest that marketization of education has become a global trend, under the tidal wave of which educational financing, curriculum, governance, and management should have been reoriented/reshaped by market-oriented approaches and practices (Currie and Newson, 1998; Taylor et al., 1997; Spring, 1998; Chan, 2001).

This chapter is set against such a wider policy context, as discussed earlier, in order to examine how educational developments have been influenced by strong market forces in post-Mao Mainland China, spearheaded by Shanghai and its Hong Kong Special Administrative Region. In particular, special attention has been given to discuss what strategies Shanghai, the so-called "dragon-head" of China's development, has adopted in response to this market economy context, in order to compare the cities of Hong Kong and Shanghai in their educational reform policies.

Conclusion: The Changing State-Education Relationships

The "marketization project" in Shanghai's education reform must be understood and analyzed in light of the wider context of the "demonopolization" of the state's role in the public domain (Mok, 1999), with its own way of educational governance. Hence, we must not analyze "marketization practices" in education in these two places simplistically in terms of a one-dimensional movement from "the state" (understood as nonmarket and bureaucratic) to "the market" (understood as nonstate and corporate).

In the case of Hong Kong, the move to market forms of production and competition in education does not constitute a withdrawal of government provisions. Instead, such a move occurs within the public sector to be accompanied by more regulations of performance. It is in this sense that Hong Kong has been instrumental in using managerialism as the cornerstone for its overall public management reforms in the public sector. On the other hand, though the move to "user pays" and the rise of nonstate provisions in Shanghai suggest a withdrawal of the state in provision and subsidy, a closer scrutiny reveals that such practices are compatible with the development of a more effective state role in terms of its governance in the "socialist market" economy. Hence, the international phenomenon of marketization in the two places should be understood differently, each in its own right. In fact, one can say that these are two different kinds of local responses to the similar global trend.

Using Shanghai as the case study to illustrate the dual processes of centralization and decentralization, it has been argued that both the territorial and functional dimensions of these two processes have occurred in China, mainly financial stringency. In this way, "central governments which realize that they do not have sufficient resources for adequate provision of services may choose to evade the problem by decentralizing responsibility to lower tiers or to nongovernmental bodies. . . . It has also been a major motive in various privatization initiatives" (Bray, 1999, p. 211).

What really happens in Shanghai is a transformation of the government's role as a sole provider of educational services to a regulator or a service purchaser. One point that deserves attention here is that during the process of transformation, the state does not "go away" in the Chinese context. Rather, the nature of the work it does has changed, very broadly speaking, "from carrying out most of the work of the coordination of education itself to determining where the work will be done and by whom" (Dale, 1997, p. 274). Seen in this light, the policy of decentralization and devolution does not necessarily mean that the state has reduced its control over the educational sphere. Instead of being merely a provider in education whereby the state has to hold the primary responsibility in service provision, the shift from direct state control to governance has caused fundamental changes in the state's role in education.

Realizing that modern states have to run their businesses with limited resources in the present social and economic context, different modes of governance have emerged in the educational sector. As for Shanghai, although the nature of the government did change in a very broad sense, what actually transformed was the government, which went "from carrying out most of the work of education itself to determining where the work will be done and by whom."

More specifically, the policy of decentralization has not entirely "deregulated" the control of the government in education. Introducing "functional decentralization" to allow individual educational institutions to have more autonomy in deciding their own development strategies, does not necessarily mean that the state has genuinely reduced its control over the education sector. Instead, the government has used other measures in supervising the various educational institutions, and has thus empowered the government to assert its direct control over the educational sector and achieve the objectives of "functional centralization" (Bray, 1999).

It is in this sense that the policy of decentralization in education does not necessarily mean that the process of deregulation is evolving; on the contrary, it may mean that there is another process of reregulation or recentralization that might be going on in the future (Whitty, 1997; Currie and Newson, 1998; Mok, 2000).

Indeed, "the Chinese state appears to be caught between a centralist, corporatist ideology (Leninism) and an economic market movement toward decentralization, which is dragging other social sectors, such as education, along" (Hawkins, 2000, p. 451).

Clearly commercial aspects, user pays, and a more expanding role for private provisions of education in Shanghai do suggest the reduced government role in educational provision and financing, but this process does not constitute a total withdrawal from governmental control and shaping. Despite the fact that the government has tried to "roll back" from the direct provider role, it has taken different forms of state intervention. Moving from direct control to governance, other actors and stakeholders in the non-state sectors have emerged to become involved in its educational provisions.

As Dale puts it, "[A]s far as coordinating institutions are concerned in relation to different governance activities in education like funding, regulation, provision and delivery, the role of the state, market and community would normally be identified" (Dale, 1997, p. 275). Here, we can see that the local government, the local communities, and the evolving market are all coordinating with each other in order to provide for new venues of educational provision.

However, as far as the processes of marketization and decentralization in Shanghai's education sector are concerned, "the state has never done all these things alone, the market and especially the community have been indispensable to the operation of the educational systems" (Dale, 1997, p. 275). In other words, one should not analyze these processes of marketization and decentralization as simply a one-dimensional shift from the "state" (as bureaucratic and nonmarket) to the "market" (as corporate and nonstate) (Mok, 2000). Because "civil society," or local communities for that matter, can play a more dominant role in between these two poles, we believe this societal restructuring will have a strong bearing upon the various levels of governance in the future.

Society is more than a sum of its individuals, but it creates the conditions within which such individuals live and develop. It is in this sense that decentralization does not necessarily mean deregulation. On the contrary, it is a process of reregulation by the government to use its resources in a more efficient and effective manner in running educational services in both Hong Kong and Shanghai. To this extent, it is crucial that we strike a delicate balance between the state and the market by empowering the often-neglected dimension and function of a "civil society" in bringing about a new vision of a society in which humans live with dignity and respect.

Whether Hong Kong or Shanghai can rise to the challenges of globalization and ride on the tide, so to speak, in order to make the best use of what it can offer will really depend upon the visions and missions that policy makers and decision makers have for their respective societies, as well as how well these policies are going to be implemented.

Table 4.1

Gross Enrollment Ratio (Primary)[a] in Shanghai

	Total number of school-age children	Total number of primary pupils	Gross enrollment ratio (%)
1993	1,090,074	1,087,109	99.73
1994	1,062,574	1,062,236	99.97
1995	1,042,821	1,042,061	99.93
1996	1,023,240	1,022,051	99.88
1997	987,972	987,686	99.97
1998	922,789	922,643	99.98
1999	835,637	835,540	99.99

Source: Department of Planning and Construction, State Educational Commission, ed., *Educational Statistics Yearbook of China* (Beijing: People's Education Press, 1996–1999).

[a]Gross enrollment ratio (Primary) refers to the ratio of the number of children (irrespective of age) enrolled in primary classes compared to the total number of children in the primary school age group of 13 or under.

Table 4.2

Gross Enrollment Ratio (Primary)[a] in Hong Kong

	Estimated population aged 6–11	Total number of primary pupils	Gross enrollment ratio (%)
1986/87	505,100	531,993	105.3
1987/88	505,300	534,309	105.7
1988/89	509,400	535,037	105.0
1989/90	512,200	534,450	104.3
1990/91	510,100	524,919	102.9
1991/92	496,300	515,938	104.0
1992/93	489,400	501,625	102.5
1993/94	481,500	485,061	100.7
1994/95	477,000	476,847	100.0
1995/96	472,800	467,718	98.9
1996/97	470,100	466,507	99.2
1997/98	467,500	461,911	98.8
1998/99	477,300	476,802	99.9
1999/2000	489,900	491,851	100.4

Source: Education Indicators for the Hong Kong School Education System 2000 Abridged Report (Hong Kong: Education Department, 2000), p. 5.

[a]Gross enrollment ratio (Primary) refers to the ratio of the number of children (irrespective of age) enrolled in primary classes compared to the total number of children in the typical primary school age group of 6 to 11.

Table 4.3

Average Class Size[a] in Shanghai Primary Schools

	Total number of primary pupils[b]	Total number of classes	Average class size
1995	1,097,784	25,408	43.21
1996	1,064,647	24,548	43.37
1997	1,024,402	23,773	43.09
1998	961,367	22,762	42.24
1999	919,878	21,747	41.05

Source: Shanghai Municipal Education Commission, ed., *Shanghai Educational Yearbook* (Shanghai: Shanghai Educational Publishing House, 1996–2000).

[a]The average class size in primary schools is the total number of day primary school pupils divided by the total number of classes in primary schools; [b]the total number of primary pupils includes those in secondary schools preparatory classes but excludes those in educational bases (*jidi*).

Table 4.4

Average Class Size[a] in Hong Kong Primary Schools

	Total number of primary pupils[b,d,e]	Total number of operating classes[c]	Average class size
1994/95	476,847	14,030	34.0
1995/96	467,718	13,912	33.6
1996/97	466,507	13,870	33.6
1997/98	461,911	14,058	32.9
1998/99	476,802	14,417	33.1
1999/2000	491,851	14,802	33.2

Source: Education Indicators for the Hong Kong School Education System 2000 Abridged Report (Hong Kong: Education Department, 2000), p. 8.

[a]The average class size in primary schools is the total number of day primary school pupils divided by the total number of operating classes in primary schools; [b]number of primary pupils and operating classes as of September of the year; [c]for combined class with primary pupils from different grades, the class size of a particular grade is the number of pupils of that grade. Each grade is counted as one-half, one-third, etc. of a class according to the number of grades in the class; [d]number of primary pupils includes those in ESF and international schools but excludes those in special schools and special classes; and [e]the total number of primary pupils includes those in preparatory classes reported by international primary school(s).

Table 4.5

Average Class Size^a in Shanghai Secondary Schools

	Total number of secondary pupils[b]		Total number of classes		Average class size	
	Junior	*Senior*	*Junior*	*Senior*	*Junior*	*Senior*
1995	595,046	128,945	12,246	2,752	48.59	46.86
1996	618,174	144,121	12,774	3,032	48.39	47.53
1997	569,291	175,046	11,987	3,661	47.49	47.81
1998	531,313	207,186	11,433	4,352	46.47	47.61
1999	534,024	232,828	11,519	4,908	46.36	47.44

Source: Shanghai Municipal Education Commission, ed., *Shanghai Educational Yearbook* (Shanghai: Shanghai Educational Publishing House, 1996–2000).

[a]The average class size in secondary schools is the total number of day secondary school pupils divided by the total number of classes in secondary schools; [b]number of secondary pupils includes those in complete (*wanquan*), senior and junior middle schools but excludes those in educational bases (*jidi*).

Table 4.6

Average Class Size^a in Hong Kong Secondary Schools

	Total number of secondary pupils[b,d]		Total number of operating classes[c]		Average class size	
	S1–S5	*S6–S7*	*S1–S5*	*S6–S7*	*S1–S5*	*S6–S7*
1994/95	406,233	51,966	10,514	1,803	38.6	28.8
1995/96	407,053	52,792	10,631	1,805	38.3	29.2
1996/97	412,553	53,105	10,577	1,829	39.0	29.0
1997/98	403,781	54,337	10,596	1,860	38.1	29.2
1998/99	399,307	56,565	10,563	1,906	37.8	29.7
1999/2000	396,217	58,248	10,455	1,944	37.8	30.0

Source: Education Indicators for the Hong Kong School Education System 2000 Abridged Report (Hong Kong: Education Department, 2000), p. 13.

[a]The average class size in secondary schools is the total number of secondary students divided by the total number of operating classes in secondary schools; [b]number of secondary students and operating classes as of September of the year; [c]for combined class with secondary students from different grades, the class size of a particular grade is the number of students of that grade. Each grade is counted as one-half, one-third, etc. of a class according to the number of grades in the class; and [d]number of secondary students includes those in ESF and international schools but excludes those in special schools and special classes.

Table 4.7

Pupil-Teacher Ratio[a] for Shanghai Primary Schools

	Total number of primary students[b]	Total number of primary school teachers	Pupil-teacher ratio
1995	1,097,784	54,517	20.13
1996	1,064,647	53,303	19.97
1997	1,024,402	51,800	19.78
1998	961,367	49,553	19.40
1999	871,620	46,810	18.62

Source: Shanghai Municipal Education Commission, ed., *Shanghai Educational Yearbook* (Shanghai: Shanghai Educational Publishing House, 1996–2000).

[a]Pupil-teacher ratio is defined as the ratio of the number of pupils enrolled to the number of teachers at the same level; [b]the total number of primary pupils includes those in secondary schools preparatory classes but excludes those in educational bases (*jidi*).

Table 4.8

Pupil-Teacher Ratio[a] for Hong Kong Primary Schools

Year	Total number of primary pupils[b,c]	Total number of primary school teachers[d,e]	Pupil-teacher ratio
1994/95	476,847	19,493	24.5
1995/96	467,718	19,710	23.7
1996/97	466,507	20,038	23.3
1997/98	461,911	20,616	22.4
1998/99	476,802	23,164	22.3
1999/2000	491,851	22,344	22.0

Source: Education Indicators for the Hong Kong School Education System 2000 Abridged Report (Hong Kong: Education Department, 2000), p. 27.

[a]Pupil-teacher ratio is defined as the ratio of the number of pupils enrolled to the number of teachers at the same level; [b]number of primary pupils as of September of the year; [c]total number of primary pupils includes those in preparatory classes reported by international primary school(s); [d]number of teachers as of October of the year; and [e]figures refer to permanent teachers only.

Table 4.9

Pupil-Teacher Ratio[a] for Shanghai Secondary Schools

	Total number of secondary pupils[b]	Total number of secondary school teachers	Pupil-teacher ratio
1995	723,991	46,519	15.56
1996	762,295	48,114	15.84
1997	744,337	48,660	15.29
1998	738,499	49,285	14.98
1999	766,852	50,270	15.25

Source: Shanghai Municipal Education Commission, ed., *Shanghai Educational Yearbook* (Shanghai: Shanghai Educational Publishing House, 1996–2000).

[a]Pupil-teacher ratio is defined as the ratio of the number of pupils enrolled to the number of teachers at the same level; [b]number of secondary pupils includes those in complete (*wanquan*) senior and junior middle schools but excludes those in educational bases (*jidi*).

Table 4.10

Pupil-Teacher Ratio[a] for Hong Kong Secondary Schools[b]

Year	Total number of secondary pupils[c]	Total number of secondary school teachers[d,e]	Pupil-teacher ratio
1994/95	458,199	22,257	20.6
1995/96	459,845	22,777	20.2
1996/97	465,658	23,077	20.2
1997/98	458,118	23,304	19.7
1998/99	455,872	24,016	19.0
1999/2000	453,465	24,453	18.5

Source: Education Indicators for the Hong Kong School Education System 2000 Abridged Report (Hong Kong: Education Department, 2000), p. 28.

[a]Pupil-teacher ratio is defined as the ratio of the number of pupils enrolled to the number of teachers at the same level; [b]secondary schools include sixth form classes; [c]number of secondary students as at September of the year; [d]number of teachers as at October of the year; and [e]figures on teachers refer to permanent teachers only.

Table 4.11

Percentage of Degree Holders Serving in Shanghai Primary Schools

Year	Total number of primary school teachers	Estimated total number of degree holders[a]	Percentage of degree holders (%)
1995	54,517	/	/
1996	53,303	/	/
1997	51,800	/	/
1998	49,553	/	/
1999	46,810	267	0.57

Source: Shanghai Municipal Education Commission, ed., *Shanghai Educational Yearbook* (Shanghai: Shanghai Educational Publishing House, 1996–2000).

[a]The estimated total number of degree holders is the total number of secondary school teachers multiplied by the percentage of degree holders.

Table 4.12

Percentage of Degree Holders[a] Serving in Hong Kong Primary Schools[b]

	Total number of primary school teachers	Total number of degree holders	Percentage of degree holders (%)
1994/95	19,493	2,247	11.53
1995/96	19,710	2,841	14.41
1996/97	20,038	3,986	19.89
1997/98	20,616	5,389	26.14
1998/99	21,364	6,660	31.17
1999/2000	21,367	8,058	37.71

Sources: Access to Information Officer, Education Department, Hong Kong, Fax Materials, Unpublished; *Education Indicators for the Hong Kong School Education System 2000 Abridged Report* (Hong Kong: Education Department, 2000).

[a]Degree holders included local degree and non-local degree; [b]figures included ordinary day schools but excluded special schools, practical schools, skills opportunity schools, English School Foundation schools and international schools.

Table 4.13

Percentage of Degree Holders Serving in Shanghai Secondary Schools

Year	Total number of secondary school teachers	Estimated total number of degree holders[a]	Percentage of degree holders (%)
1995	46,519	18,356	39.46
1996	48,114	19,852	41.26
1997	48,660	20,998	43.15
1998	49,285	22,672	46.00
1999	50,270	25,043	49.82

Source: Shanghai Municipal Education Commission, ed., *Shanghai Educational Yearbook* (Shanghai: Shanghai Educational Publishing House, 1996–2000).

[a]The estimated total number of degree holders is the total number of secondary school teachers multiplied by the percentage of degree holders.

Table 4.14

Percentage of Degree Holders[a] Serving in Hong Kong Secondary Schools[b]

Year	Total number of secondary school teachers	Total number of degree holders	Percentage of degree holders (%)
1994/95	22,257	15,955	71.69
1995/96	22,777	16,709	73.36
1996/97	23,077	17,615	76.33
1997/98	23,304	18,405	78.98
1998/99	24,016	19,312	80.41
1999/2000	23,434	19,959	85.17

Sources: Access to Information Officer, Education Department, Hong Kong, Fax Materials, Unpublished; *Education Indicators for the Hong Kong School Education System 2000 Abridged Report* (Hong Kong: Education Department, 2000).

[a]Degree holders included local degree and non-local degree; [b]figures included ordinary day schools but excluded special schools, practical schools, skills opportunity schools, English School Foundation schools and international schools.

Table 4.15

Government Expenditure per Primary Pupil per Year[a] in Shanghai

	Estimated government expenditure on primary education[b] (million yuan)	Total number of primary pupils[c]	Government expenditure per primary pupil (yuan)
1995	1,230.80	1,097,784	1,121.17
1996	1,499.12	1,064,647	1,408.09
1997	1,699.80	1,024,402	1,659.31
1998	1,863.73	961,367	1,938.62
1999	2,153.51	919,878	2,341.08

Source: Shanghai Municipal Education Commission, ed., *Shanghai Educational Yearbook* (Shanghai: Shanghai Educational Publishing House, 1996–2000).

[a]Government expenditure per primary pupil refers to financial budget (*caizheng bokuan*) of government per primary pupil but excludes other expenditure per primary pupil from various forms of revenue; [b]the estimated government expenditure on primary education is the total number of primary pupils multiplied by the government expenditure per primary pupil; and [c]the total number of primary pupils includes those in secondary school preparatory classes but excludes those in educational bases (*jidi*).

Table 4.16

Government Expenditure per Primary Pupil per Year[a] in Hong Kong

	Government expenditure on primary education ($ 1,000)	Total number of primary pupils[b]	Government expenditure per primary pupil ($)
1994/95	5,832.1	476,847	12,231
1995/96	6,494.4	467,718	13,885
1996/97	6,980.6	466,507	14,964
1997/98	7,595.5	461,911	16,444
1998/99	8,573.4	476,802	17,981
1999/2000	9,405.0	491,851	19,122

Source: Education Indicators for the Hong Kong School Education System 2000 Abridged Report (Hong Kong: Education Department, 2000).

[a]Government expenditure on primary education refers to the recurrent and non-recurrent expenditure of the government on primary education but excludes capital expenditure chargeable to Capital Works Reserve Fund; [b]the total number of primary pupils includes those in preparatory classes reported by international primary school(s).

Table 4.17

Government Expenditure per Secondary Pupil per Year[a] in Shanghai

	Estimated government expenditure on secondary education[b] (million yuan)	Total number of secondary pupils[c]	Government expenditure per secondary pupil (yuan)
1995	1,096.66	723,991	1,514.74
1996	1,454.09	762,295	1,907.51
1997	1,739.30	744,337	2,336.71
1998	2,068.56	738,499	2,801.03
1999	2,333.02	766,852	3,042.33

Source: Shanghai Municipal Education Commission, ed., *Shanghai Educational Yearbook* (Shanghai: Shanghai Educational Publishing House, 1996–2000).

[a]Government expenditure per secondary pupil refers to financial budget (*caizheng bokuan*) of government per secondary pupil but excludes other expenditure per secondary pupil from various forms of revenue; [b]the estimated government expenditure on secondary education is the total number of secondary pupils multiplied by the government expenditure per secondary pupil; and [c]number of secondary pupils includes those in complete (*wanquan*), senior and junior middle schools but excludes those in educational bases (*jidi*).

Table 4.18

Government Expenditure per Secondary Pupil per Year[a] in Hong Kong

	Government expenditure on secondary education ($ 1,000)	Total number of secondary pupils	Government expenditure per primary pupil ($)
1994/95	8,572.3	458,199	18,709
1995/96	9,802.5	459,845	21,317
1996/97	10,867.6	465,658	23,338
1997/98	11,921.7	458,118	26,023
1998/99	13,385.7	455,872	29,363
1999/2000	14,445.3	453,465	31,855

Source: Education Indicators for the Hong Kong School Education System 2000 Abridged Report (Hong Kong: Education Department, 2000).

[a]Government expenditure on secondary education refers to the recurrent and non-recurrent expenditure of the government on secondary education but excludes capital expenditure chargeable to Capital Works Reserve Fund.

Table 4.19

Ratio of Government Expenditure on Education[a] to GDP in Shanghai

	Total expenditure on GDP (million yuan)	Government expenditure on education (million yuan)	Expenditure on education as % of GDP
1995	246,257	3,552.40	1.44
1996	290,520	4,419.28	1.52
1997	336,021	5,290.09	1.57
1998	368,820	6,001.66	1.63
1999	403,496	7,037.31	1.74

Sources: Shanghai Municipal Statistical Bureau, ed., *Statistical Yearbook of Shanghai* (Shanghai: China Statistical Press, 1998–2000); Shanghai Municipal Education Commission, ed., *Shanghai Educational Yearbook* (Shanghai: Shanghai Educational Publishing House, 1996–2000).

[a]Government expenditure on education refers to financial budget (*caizheng bokuan*) of government on education but excludes other expenditure on education from various forms of revenue.

Table 4.20

Ratio of Government Expenditure on Education[a] to GDP[b] in Hong Kong

Financial year	Total expenditure on GDP[c] ($ 1,000)	Government expenditure on education ($ 1,000)	Expenditure on education as % of GDP
1994/95	1,010,885	28,528.0	2.8
1995/96	1,077,145	32,816.7	3.0
1996/97	1,191,890	36,764.3	3.1
1997/98	1,325,165	41,661.2	3.1
1998/99	1,262,296	47,230.3	3.7
1999/2000	1,227,658	52,252.0	4.2[d]
2000/01	1,267,175	54,383.0	4.1[e]

Sources: Education Indicators for the Hong Kong School Education System 2000 Abridged Report (Hong Kong: Education Department, 2000); Census and Statistics Department, Hong Kong, *Hong Kong Statistics*, available at: www.info.gov.hk/censtatd/eng/hkstat/fas/nat_account/gdp/gdp1_index.html; Access to Information Officer, Education Department, Hong Kong, Mail Materials, Unpublished.

[a]Government expenditure on education includes education subventions and expenditure of the Education Department, Hong Kong Institute of Vocational Education (on technical and vocational education), and University Grants Committee (on tertiary education); [b]Gross Domestic Product (GDP), a concept very similar to Gross National Product (GNP), is equal to the total of the gross expenditure on the final uses of the domestic supply of goods and services valued at price to the purchaser minus the imports of goods and services. The total expenditure figures on GDP are expenditure-based estimates at current market prices; [c]total expenditure on GDP is expressed at current market prices; [d]revised estimates by Education and Manpower Bureau, Government Secretariat; and [e]draft estimates by Education and Manpower Bureau, Government Secretariat.

Table 4.21

Government Expenditure on Education[a] as a Percentage of Total Government Expenditure in Shanghai

	Local expenditure[b] (million yuan)	Government expenditure on education (million yuan)	Percentage of expenditure on education
1995	26,789	3,552.40	13.26
1996	34,266	4,419.28	12.89
1997	42,892	5,290.09	12.33
1998	48,070	6,001.66	12.49
1999	54,638	7,037.31	12.88

Sources: Shanghai Municipal Statistical Bureau, ed., *Statistical Yearbook of Shanghai* (Shanghai: China Statistical Press, 1998–2000); Shanghai Municipal Educational Commission, ed., *Shanghai Educational Yearbook* (Shanghai: Shanghai Educational Publishing House, 2000).

[a]Government expenditure on education refers to financial budget (*caizheng bokuan*) of government on education but excludes other expenditure on education from various forms of revenue; [b]local expenditure includes expenditure on capital construction; technical updates and transformoction of enterprises; supporting agricultural production; city maintenance; culture, education, science, and health care; administration; and science and technology promotion.

Table 4.22

Government Expenditure on Education[a] as a Percentage of Total Government Expenditure[b] in Hong Kong

	Total government expenditure ($ 1,000)	Government expenditure on education ($ 1,000)	Percentage of expenditure on education
1994/95	164,155.3	28,528.0	17.4
1995/96	183,158.0	32,816.7	17.9
1996/97	182,680.0	36,764.3	20.1
1997/98	194,360.0	41,661.2	21.4
1998/99	239,356.0	47,230.3	19.7
1999/2000	223,043.2	52,252.0	23.4 (18.8)[c]
2000/01	232,893.6	54,383.0	23.4 (18.9)[d]

Sources: Education Indicators for the Hong Kong School Education System 2000 Abridged Report (Hong Kong: Education Department, 2000); Access to Information Officer, Education Department, Hong Kong, Mail Materials, Unpublished; Treasury, Government of the Hong Kong SAR, 2000–2001 Annual Account, available at: www.info.gov.hk/tsy/intrnet/html/statend.html.

[a]Government expenditure on education includes education subventions and expenditure of the Education Department, Hong Kong Institute of Vocational Education (on technical and vocational education), and University Grants Committee (on tertiary education); [b]the total government expenditure includes operating expenditure, capital expenditure, and equity investments; [c]number in parentheses is a revised estimate by Education and Manpower Bureau, Government Secretariat; and [d]number in parentheses is a draft estimate by Education and Manpower Bureau, Government Secretariat.

Table 4.23

Distribution of Education Attainment of Population Age Fifteen and Over in Shanghai

	Illiterate and semi-illiterate		Primary[a]		Junior secondary[a]		Senior secondary		College and higher level		Total	
	M	F	M	F	M	F	M	F	M	F	M	F
1996	3.8	15.4	15.2	16.2	38.9	33.9	27.2	25.7	14.9	8.8	100	100
1997	4.2	15.9	15.5	17.1	39.2	33.4	27.8	26.5	13.3	7.1	100	100
1998	4.3	15.9	16.5	16.8	36.4	32.1	28.7	26.9	14.1	8.3	100	100
1999	3.5	13.9	13.7	16.0	37.3	32.8	30.3	27.9	15.4	9.4	100	100

Sources: Department of Population and Employment Statistics, State Statistical Bureau, ed., *China Population Statistics Yearbook* (Beijing: China Statistics Press, 1997–2000); Department of Planning and Construction, State Educational Commission, ed., *Educational Statistics Yearbook of China* (Beijing: People's Education Press, 1996–1999).

[a]The percentages of primary and junior secondary are estimated according to the total number of primary pupils of that year.

Table 4.24

Distribution of Education Attainment of Population Age Fifteen and Over in Hong Kong

	No Schooling/ kindergarten		Primary		Secondary/ matriculation		Tertiary (non-degree)		Tertiary (degree)		Total	
	M	F	M	F	M	F	M	F	M	F	M	F
1995	12.6	20.2	28.7	26.7	45.8	42.3	5.1	5.3	7.8	5.5	100	100
1996	11.6	18.9	28.0	26.2	46.4	43.2	5.3	5.7	8.7	6.0	100	100
1997	11.4	18.1	27.2	25.8	46.1	43.3	5.9	6.1	9.4	6.7	100	100
1998	10.9	17.3	27.3	25.9	46.4	43.7	5.6	5.9	9.8	7.2	100	100
1999	10.3	16.4	26.7	26.0	47.3	44.1	5.7	5.7	10.0	7.8	100	100
2000	9.5	15.2	26.7	26.0	47.2	44.6	5.9	5.9	10.7	8.3	100	100

Source: Access to Information Officer, Education Department, Hong Kong, Mail Materials, Unpublished.

Note: Figures from the second quarter 1996 onward have taken into account all the recent technical enhancements in the estimation method for population statistics and the aggregate results of the 2001 Population Census.

Table 4.25

Comparison of Hong Kong and Shanghai Education Reform Directions

	Hong Kong	Shanghai
Main features:	Decentralization in schools Marketization shifted toward corporate management approach Adoption of principles of managerialism: efficiency, effectiveness, and economy Privatization as an instrument of economic and social policies in times of financial stringency	Changing national policy context in the Mainland Decentralization to local authorities Making use of market forces and new initiatives from non-state sectors Marketization and privatization of educational services
Merits:	9 years of compulsory basic education More opportunities and equitable access to basic, tertiary, and continuing education Schools and universities are competitive on the basis of performance indicators Parents have more choices and clearer standards in selecting schools and universities for their children	9 years of free basic education More opportunities and equitable access to basic, tertiary, and continuing education Schools and universities are increasingly competitive on the basis of performance indicators Parents have more choices and clearer standards in selecting schools and universities for their children
Problems:	As the management of schools and universities is more market-driven, the curricula will tend to be market-oriented, resulting in the possible reductions or losses of subject matters, such as philosophy or history, that are of no significant market value and demand Accountability and intensified work pressures for educational practitioners	Decentralization in terms of financial resources causes disparity in educational financing Disparity in educational developments within Greater Shanghai
Reform directions:	Call for quality education and lifelong learning to meet the challenges of knowledge-based global economy Principles of education reforms: (1) student-focused, (2) "no-loser" in the system, (3) quality education, (4) life-wide learning, and (5) society-wide mobilization	Developmental goal of Shanghai as the "national city" with first-class education to become the international economic, financial, and trade center Territorial decentralization Functional centralization Increasing role of local government in higher education

(*continued*)

Table 4.25 *(continued)*

Hong Kong	Shanghai
Main focuses are to: (1) reform the admission systems and public examinations, (2) reform the curricula and improve teaching methods, (3) improve the assessment mechanism, (4) provide lifelong learning at various levels, (5) formulate an effective resource strategy, (6) enhance the professionalism of teachers, and (7) implement measures to support frontline educators Self-financing programs Diversified modes in educational delivery Budget cuts across all educational sectors	Emergence of community/society-wide education Establishing more international links with overseas higher education institutions

—— 5 ——

Labor

Grace Lee, Wang Daben, and Simon Li

Hong Kong and Shanghai have followed different political, social, and economic systems. Hong Kong has always been popular for its free economy while China has leaned toward an alleged socialist market economy. However, in employment services, the two places have shared some resemblance in that there has been a shift from manufacturing to servicing industries, as well as a rise in unemployment and the existence of active labor market programs as important employment policy tools.

This chapter examines the employment policies and active labor market programs in the two urban cities of China in spite of their diversities and brings out a comparison on how these two places have dealt with unemployment problems. The chapter begins by comparing the economic growth of Hong Kong and Shanghai, followed by a section on a statistical overview of the labor market conditions and trends in employment. Then the chapter proceeds to analyze the dissimilarities and similarities in the causes of unemployment and the active labor market programs taken up by the two governments.

Comparing Hong Kong's and Shanghai's Economic Growth

Both Shanghai and Hong Kong share similar characteristics of rapid economic growth. Beginning in late 1978 the Chinese leadership has been moving the economy from a sluggish Soviet-style centrally planned economy to a more market-oriented economy, but still within a rigid political framework of Communist Party control. To this end the authorities have switched to a system of household responsibility in agriculture in place of the old collectivization, increased the authority of local officials and plant managers in industry, permitted a wide variety of small-scale enterprise in services and light manufacturing, and opened the economy to increased foreign trade and investment. The result has been a quadrupling of gross domestic product (GDP) since 1978. In 2000, with its 1.26 billion people and a

Figure 5.1 **GDP Growth of China and Shanghai, 1980–2000**

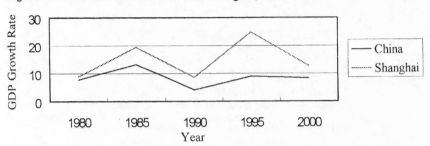

Sources: State Statistical Bureau, *Zhongguo Tongji Nianjian 2001* (China Statistical Year-
book 2001) (Beijing: Zhongguo Tongji Chubanshe), pp. 49–50; Shanghai Statistical Bureau,
Shanghai Tongji Nianjian 2001 (Shanghai Statistical Yearbook 2001) (Shanghai: Shanghai
Tongji Chubanshe), pp. 328–29.

GDP of just $3,600 per capita, China stood as the second-largest economy in the
world after the United States (Central Intelligence Agency, 2001).

Rapid economic growth began in the major cities along the eastern coast of
China, and Shanghai was among the fastest-growing economies. Figure 5.1 shows
that its economy outgrew the national average in the last decade or so.

On the other hand, Hong Kong was described as having "a bustling free market
economy highly dependent on international trade. Natural resources are limited,
and food and raw materials must be imported. Indeed, imports and exports, including re-exports, each exceed GDP in dollar value. . . . Per capita GDP compares
with the level in the four big countries of Western Europe. GDP growth averaged
a strong percent in 1989–97" (Central Intelligence Agency, 2001). In fact, over the
past two decades, the Hong Kong economy has almost tripled, with the GDP growing at an average annual rate of about 5 percent in real terms. Per capita GDP in
Hong Kong has more than doubled in real terms, equivalent to an average annual
growth rate of about 4 percent in real terms. In 1999, it reached U.S.$23,200 at
current market prices (Hong Kong Annual Report, 2000).

Although both Shanghai and Hong Kong share similar characteristics of rapid
economic growth, their economic structure is drastically different. "Chinese Marxist
and statist theories alike place a high value on one service: government. Other
services have received less consistent support. Whatever rational justification might
be offered for commerce, insurance, finance, tourism, brokerage, or advertising, in
Mao's time radicals deemed these activities 'non-productive,' frivolous at best and
exploitative at worst" (White, 1998, p. 433). Table 5.1 shows that up until 2000,
71 percent of employment in Shanghai city was absorbed by state-owned enterprises and collective enterprises, leaving only 29 percent with other economic units.

On the contrary, public expenditure in Hong Kong occupied a relatively small

Table 5.1

Composition of Urban Employment in Shanghai

		Composition of employment (%)		
Year	Number of employees ('000)	State-owned enterprises	Collective enterprises	Others
1980	4,469.2	78.7	21.3	–
1990	5,081.0	78.2	20.0	1.8
2000	3,901.4	58.8	12.2	29.0

Source: *Shanghai Tongji Nianjian 2001* (Shanghai Tongji Chubanshe 2001).

proportion of 22.6 percent of its GDP (Hong Kong Annual Report, 2000, Appendix 4). As a result of the marked difference in economic structure between Shanghai and Hong Kong, the challenges to employment differ in the two societies. The following section is a statistical overview of the labor market conditions and trends in employment.

Statistical Overview of the Labor Market Conditions and Trends

As of July 2001, Hong Kong had a population of 7.210 million. Its labor force stood at 3.4 million (2000 estimate). The population of Shanghai was 16.74 million in 2000, of which 7.5 million constituted its labor force. The relative importance of the various economic sectors can be gauged by their respective contributions to GDP and total employment (see Table 5.2).

Structural shifts from the manufacturing industry to the servicing sector could be traced in both cities. Table 5.4 shows a change in the number of persons engaged in the different sectors in Hong Kong and Shanghai between 1980 and 2000.

In 2000, 86.4 percent of the workers in Hong Kong were engaged in the servicing sector as compared to 44.9 percent in Shanghai. Figure 5.1 shows that the number in the servicing sector has risen in both places though in different proportions.

In Hong Kong, owing to the absence of natural resources, primary production is very small in terms of its contributions to both GDP and employment. Within secondary production, the contribution of the manufacturing sector to GDP fell significantly over the past two decades. This was mainly due to the combined influence of a fast expansion in the services sector in Hong Kong and the ongoing relocation of the more labor-intensive manufacturing processes to the Mainland. A structural shift during the past decade has meant establishments in the service sec-

Table 5.2

GDP and Employment by Broad Economic Sectors in Shanghai and Hong Kong

| | Shanghai | | | | | | Hong Kong | | | |
| | Percentage of GDP (%) | | | Percentage of labor force (%) | | | Percentage of GDP (%) | | Percentage of labor force (%) | |
	1978	1990	2000	1978	1990	2000	1980	1999	1980	2000
Primary industries	4.0	2.3	1.8	34.5	11.1	10.8	1.0	0.1	1.5	0.4
Secondary industries	77.4	63.8	47.6	11.1	59.3	44.3	31.6	14.5	50.1	17.3
Tertiary industries	18.6	31.9	50.6	21.4	29.6	44.9	67.5	85.4	48.4	82.3
Total (%)	100	100	100	100	100	100	100	100	100	100

Sources: Shanghai Tongji Nianjian 2001 (Shanghai Tongji Chubanshe, 2001); *Hong Kong Annual Report 2000*, Hong Kong SAR government.

tors now employ about eight times as many workers as the manufacturing sector. In 2000, 2,022,500 persons were engaged in establishments in the various service sectors (not including most of the self-employed and those engaged in the provision of personal services), 102,500 persons more than the corresponding figure in 1999. Only 229,400 persons were engaged in manufacturing sector establishments (excluding outworkers).

In Shanghai, secondary and primary productions contributed mainly toward the GDP and employment in the 1980s as compared with tertiary production. However, in the 1990s there was a shift of the primary production to smaller areas and the secondary and tertiary contributed mainly to both GDP and employment. "Output gains, at an early stage of modern growth, involve the migration of ex-agricultural wage labor into industry—and largely into towns. These workers create manufactured products at traditional low subsistence pay (plus a nominal mark up) as long as the supply of them lasts. The value of what they make is greater than what they receive, and the capital-accumulating network takes the difference" (White, 1998, pp. 407–8). "Half the migrants to Shanghai suburban towns in the 1980s admitted they came 'to change' jobs. People left rural China in droves to seek employment, in cities as large as Shanghai if they could" (p. 418). "Two-thirds of Shanghai's 'agricultural' workers (over three million people) by the late 1980s were actually employed in suburban industries and services. Municipal officers publicly reckoned that the city then had 'several hundred thousand' contract workers. The vagueness of the estimates suggests that nobody was able to make a reliable count. Village factory managers, protected by local cadres against any effective state controls, had many reasons not to divulge their full list of workers" (p. 420). The influx of people into Shanghai brought with it the problem of rising unemployment. These migrant workers from poorer regions of the country used to provide many of these services as the local "labor aristocrats" often felt too ashamed to mention these jobs to their family and friends. Yet, gradual reeducation has helped change the workers' mentality, and they have become more receptive to the idea of service provision. In 2000, 3,720,800 persons were engaged in establishments in the various service sectors, 3,670,400 in the secondary production, and only 892,300 in the primary production.

Unemployment in Hong Kong and Shanghai

Since the 1980s, much of Hong Kong's manufacturing has shifted out of Hong Kong proper in search of lower-cost land and labor. Hong Kong enterprises now own and manage a far-reaching network of activities in the Pearl River Delta farther inland in China and beyond. The scale of Hong Kong's manufacturing activities has greatly expanded, even as production within Hong Kong itself has shrunk (Berger and Lester, 1997, p. xii). A team of researchers from the Massachusetts Institute of Technology coined the term "Made by Hong Kong" to refer to the products manufactured by Hong Kong–owned and/or managed enterprises wherever they are located. It designates in a phrase the international chains of networked production that link Hong Kong–owned and –managed companies, inside and outside of Hong Kong proper, whereas "Made in Hong Kong" refers to goods and services that are produced within the territory of Hong Kong proper. Many of these goods and services serve as inputs into the production networks of "Made by Hong Kong." In brief, "Made *by* Hong Kong" has grown, even as "Made *in* Hong Kong" has declined (p. xii). The Hong Kong "drama" of "deindustrialization" in the later half of the eighties and early nineties has yielded a labor surplus of almost half a million factory workers. They were released from factories, which were located, or relocated, en masse northward to the Mainland. Nevertheless, the bulk of this dislocated industrial workforce was largely absorbed by the booming economy and growth in tertiary industries, hence explaining the low unemployment rate before Hong Kong was hit by the Asian financial crisis in 1997 (see Figure 5.2).

The economy of Hong Kong was badly hit by the Asian financial crisis. The property market and the stock market plunged. Figures released by the Land Registry showed 55,178 transactions recorded in the first half of 1998, down 52.3 percent from the same period in 1997. The figure is a 38.5 percent decline from the second half of 1997. The value of the first half-year transactions in 1998 amounted to $183.97 billion, down 62 percent from last year's first half and 52.1 percent from the second half (Ko, 1998). The Hang Seng Index plunged from

Figure 5.2 **Unemployment Rate of Hong Kong, 1980–2000 (in %)**

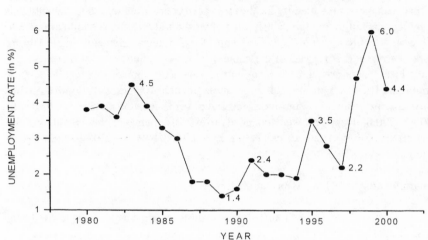

16,800 to around 8,000 in August 1998. Trade and tourism were badly hit as Hong Kong's trade partners were suffering from an even more severe setback.

The Asian financial crisis has precipitated a wave of business collapses and closures that inevitably led to layoffs and retrenchment. Table 5.3 shows incidents of mass layoffs after the crisis.

Parallel to business closures is the emergence of austerity measures adopted elsewhere in the private sector to rationalize, consolidate, and cut back in labor costs. Downsizing, flexi-hiring, flexi-pay package, pay pause, outsourcing, short-time working, job sharing, extended holiday and leave of absence, part-time and piecework hiring, as well as other associated practices of labor-saving measures have become common practice not only in small industries but also in the primary sector (Ng and Lee, 1998, p. 173). Table 5.4 shows the results of a survey conducted by the Hong Kong Institute of Human Resource Management in early 1999 on measures to cope with the problem of redundancies adopted by its member organizations.

According to the labor force statistics by the Census and Statistics Department, the seasonally adjusted unemployment rate in July–September 2001, at 5.3 percent (provisional), was higher than that in June–August 2001, at 4.9 percent. The underemployment rate in July–September 2001, at 2.5 percent (provisional), was also higher than that of 2.2 percent in June–August 2001. Comparing July–September with June–August 2001, increases in the unemployment rate (not seasonally adjusted) occurred mainly in the construction, manufacturing, restaurants, import/export trades, and retail trade sectors. Affected by the Asian financial crisis, the unemployment rate in Hong Kong reached a record height of 6.3 percent in February–April 1999. Altogether, 216,000 persons were out of jobs. The situation improved in 2000 when the unemployment rate dropped to 4.4 percent. However,

Table 5.3

Impact of the Asian Financial Crisis on Employment in Hong Kong

Company	Business	Impact
Theme International Holdings	Retail	Closed 23 stores, sacked 86 workers
G2000	Retail	Cut wages of 600 staff
Hong Kong Telecom	Telecommunications	270 managers and technicians sacked
Wharf Cable	Telecommunications	"Several dozen" staff in administration and human resources departments made redundant
Tin Tin Daily News	Newspaper	Staff salary cut of up to 20 percent
ATV	Television	135 staff sacked
Furama Hotel	Hotel	40 workers sacked
Jardine Flemming	Investment	40 employees dismissed
Daimaru	Department store	Closed 2 department stores, 400 jobs lost
Wing On Department Store	Department store	40 workers laid off, 270 employees made redundant
Miramar Hotel & Investment	Hotel	Hotel's workforce reduced by 15 percent through natural wastage
Cathay Pacific Airways	Airline	About 1,870 jobs lost
The Excelsior	Hotel	Hundreds of workers forced to accept new contracts
Daiwa Securities (HK)	Securities	More than a third of its staff culled
First Pacific Bank	Banking	40 staff sacked
First Pacific Davies	Property	47 employees fired
HSBC James Capel Asia	Finance	1 in 12 of its staff laid off
Scott Wilson (HK)	Engineering design	15 staff made redundant
Peregine's	Finance	700 staff laid off

Source: South China Morning Post, Hong Kong.

2001 experienced a high rate of unemployment. The latest figures show that the jobless rate for the September–November 2001 period climbed to 5.8 percent. More than 60,000 jobs had been lost in the past year with 205,000 people out of work.

The number of vacancies shrank from 47,200 in the first quarter of 1997 to 25,350 in the fourth quarter of 1998. Hong Kong has never experienced such a serious unemployment situation for more than three decades. A high-level Task

Table 5.4

Organizational Measures Coping with Redundancy in Hong Kong (January–June 1999)

Methods used to cope with excessive manpower	Top/Senior management	Middle management	Non-Management professional	Supervisory/Officer	Technical	Clerical/Secretarial	Junior/Manual/frontline	All staff
1. Froze recruitment	35.6%	47.5%	49.2%	49.2%	35.6%	59.3%	52.5%	76.3%
2. Natural wastage	33.9%	39.0%	42.4%	42.4%	30.5%	55.9%	52.5%	69.5%
3. Early retirement	10.2%	5.1%	5.1%	8.5%	6.8%	8.5%	6.8%	20.3%
4. Redundancy								
a. Voluntary	3.4%	8.5%	6.8%	5.1%	5.1%	5.1%	6.8%	11.9%
b. Involuntary	5.1%	32.2%	27.1%	37.3%	25.4%	49.2%	35.6%	62.7%
i. By performance	1.7%	15.3%	11.9%	22.0%	16.9%	28.8%	23.7%	40.7%
ii. By length of service	0.0%	3.4%	3.4%	5.1%	3.4%	5.1%	3.4%	6.8%
iii. By age	0.0%	1.7%	0.0%	0.0%	0.0%	0.0%	0.0%	1.7%
iv. By job	3.4%	23.7%	18.6%	25.4%	16.9%	32.2%	27.1%	44.1%
5. Changed permanent staff/position to contract-based	0.0%	1.7%	3.4%	8.5%	8.5%	5.1%	10.2%	20.3%
6. Redeployment to subsidiaries or other associated companies	5.1%	10.2%	8.5%	11.9%	8.5%	11.9%	8.5%	15.3%
7. Outsourced certain functions	1.7%	5.1%	5.1%	3.4%	6.8%	6.8%	11.9%	20.3%
8. Other	3.4%	5.1%	3.4%	5.1%	3.4%	5.1%	6.8%	6.8%

Source: Institute of Human Resource Management, "The Effectiveness of HR Strategies on Business Performance since the Financial Crisis," in *Human Resources*, Vol. 4, 11 (Hong Kong: Institute of Human Resource Management, 1999), pp. 14–16.

Force on Employment chaired personally by the financial secretary of the Hong Kong government met monthly starting in mid-1998 to promote job creation and to monitor the employment situation. The public employment service was under unprecedented pressure to innovate and to deliver. Yet, closer examination of the profile of the unemployed, paying particular attention to their previous employment by industry and by occupation, will reveal the fact that Hong Kong has not only transited from the manufacturing to the servicing era, but also has moved into a knowledge-based economy (see Table 5.5). The rapid rise of the knowledge-based economy led to a great demand for well-trained talent with the specialized skills required for technological development and application. However, most low-income workers do not have the necessary skills or the educational background to adapt to the new economy.

In Shanghai, the unemployment rate of 3.5 percent in 2000 was lower than that in Hong Kong (4.4 percent). However, caution has to be paid to differences in definitions. The definition of unemployment is distinctive in China. According to the State Statistical Bureau, unemployment refers to the urban registered unemployed who (1) possess a nonagricultural residence, (2) are within a certain age range (16 to 50 for males and 16 to 45 for females), (3) are able and willing to work, and (4) have registered with the local labor bureau for employment (*Chinese Labor Statistical Yearbook*, 1997, p. 588). Only the openly unemployed are eligible for unemployment benefits. In fact, another form of unemployment is perhaps more pervasive—"hidden unemployment"—referring to workers, often in the state sector, who have been laid off (*xiagang*). The State Statistical Bureau defines laid-off workers to be "workers who have left their posts and are not engaged in other types of work in the same unit, but still maintain a labor relationship with the unit that they have worked" (p. 588). Workers who have been laid off are only given living subsidies (*shenghuofei*) instead of unemployment benefits, and are not included in the registered unemployment rate.

Hong Kong adopts the definition of unemployment as laid out by the International Labor Organization, that is, for a person aged fifteen or over to be classified as unemployed, that person should (1) not have had a job and should not have performed any work for pay or profit during the seven days before enumeration, (2) have been available for work during the seven days before enumeration, and (3) have sought work during the thirty days before enumeration (Commissioner for Labour, 2000, p. 76). Moreover, the concept of "underemployment" does not exist in China. In the case of Hong Kong, the criteria for an employed person to be classified as underemployed are involuntarily working less than 35 hours during the seven days before enumeration or has sought additional work during the thirty days before enumeration (p. 76).

An examination of the causes of unemployment in Shanghai, hidden and open, reveals similar structural shifts in employment (see Table 5.6).

Shanghai fell into the footsteps of Hong Kong in the 1980s and 1990s, namely, in the decline of the manufacturing sector, albeit to a lesser extent. A more

Table 5.5

Profile of the Unemployed in Hong Kong (Third Quarter, 2001)

	Male		Female		Overall	
Age	*No. ('000)*	*Rate* (%)*	*No. ('000)*	*Rate* (%)*	*No. ('000)*	*Rate* (%)*
15–19	10.4	27.2	7.8	22.8	18.2	25.1
20–29	36.2	9.1	24.3	5.7	60.5	7.4
30–39	23.9	4.3	11.6	2.4	35.5	3.4
40–49	33.1	5.7	12.0	3.3	45.1	4.8
50–59	18.5	5.9	4.9	3.6	23.4	5.2
≥60	2.7	2.9	0.2	0.9	2.9	2.5
Education level	*No. ('000)*	*Rate* (%)*	*No. ('000)*	*Rate* (%)*	*No. ('000)*	*Rate* (%)*
No schooling/kindergarten	1.8	7.9	0.5	1.8	2.2	4.7
Primary	26.0	7.8	7.6	3.7	33.6	6.2
Secondary/matriculation	78.6	6.9	37.9	4.4	116.5	5.8
Tertiary						
-non-degree	7.5	5.0	6.5	4.5	14.0	4.7
-degree	11.0	3.4	8.3	3.6	19.3	3.5
Previous job by industry	*No. ('000)*	*Percentage (%)*	*No. ('000)*	*Percentage (%)*	*No. ('000)*	*Percentage (%)*
Manufacturing	12.9	8.1	6.9	4.3	19.8	12.4
Construction	32.5	20.3	1.3	0.8	33.8	21.2
Wholesale, retail and import/export trades, restaurants, and hotels	34.3	21.5	26.1	16.3	60.4	37.8
Transport, storage, and communications	11.8	7.4	2.1	1.3	13.9	8.7
Financing, insurance, real estate, and business services	10.1	6.3	5.1	3.2	15.2	9.5
Community, social, and personal services	7.9	4.9	8.4	5.3	16.3	10.2
Others	0.4	0.2	0	0	0.4	0.2
Previous job by occupation	*No. ('000)*	*Percentage (%)*	*No. ('000)*	*Percentage (%)*	*No. ('000)*	*Percentage (%)*
Managers and administrators	4.7	3.0	0.9	0.6	5.6	3.5
Professionals	2.8	1.7	1.6	1.0	4.4	2.8

(continued)

Table 5.5 *(continued)*

	Male		Female		Overall	
Age	No. ('000)	Percentage* (%)	No. ('000)	Percentage* (%)	No. ('000)	Percentage* (%)
Associate professionals	11.6	7.3	6.3	4.0	17.9	11.2
Clerks	6.4	4.0	14.9	9.3	21.3	13.3
Service workers and shop sale workers	19.1	12.0	16.0	10.0	35.1	22.0
Craft and related workers	31.0	19.4	0.6	0.4	31.6	19.8
Plant and machine operators and assemblers	8.8	5.5	3.8	2.4	12.6	7.9
Elementary occupations	25.2	15.8	5.9	3.7	31.1	19.5
Others	0.1	0.1	0	0	0.1	0.1

Source: Quarterly Report on General Household Survey (July–September 2001), Census and Statistics Department, Hong Kong SAR government.

*Unemployment rate in respect of the specified sex/age group.

fundamental factor leading to the structural shift and employment situation in Shanghai could be attributed to the macrostructural change from command economy to market economy in China. In the Mao Zedong era, labor allocation was rigid. The young school leavers were assigned to a work unit (*danwei*), which registered their citizenship status (*hukou*) (see Cheng and Selden, 1994). The mismatching of jobs to people often resulted in discontentment and subsequently poor productivity and low morale. As Shenkar and von Glinow stated, "[H]owever disgruntled they may be, employees are reluctant to leave the work unit to which they have been assigned out of fear of remaining unaffiliated in a society in which major necessities are supplied only through organizational affiliation" (Shenkar and von Glinow, 1994, p. 61). The enterprises run the society (*qiye ban shehui*). The *danwei* constituted a mini–welfare state, providing not only jobs and earnings but also a wide array of goods and services for employees and their families, ranging from housing, medical care, educational provisions, and child care to pensions and crematorial service. Most workers and staff on the state payroll had lifetime tenure in their original units with job promotions or wage increments based heavily on seniority.

Overstaffing has long been common in a labor surplus economy like China's. From the mid-1950s on, the Chinese party-state has taken upon itself the task of finding jobs for the vast majority of urban job seekers. In consequence, particularly

Table 5.6

Unemployment in Hong Kong and Shanghai

	Hong Kong	Shanghai
Official unemployment rate	7.8% (2002)	3.5% (2001)
Definition of unemployment	For a person aged 15 or over to be classified as unemployed, that person should (a) not have had a job and should not have performed any work for pay or profit during the seven days before enumeration; and (b) have been available for work during the seven days before enumeration; and (c) have sought work during the thirty days before enumeration (International Labour Organization)	Urban registered unemployed who (a) possess non-agricultural residence; (b) are within a certain age range (16 to 50 for males and 16 to 45 for females); (c) are able and willing to work; and (d) have registered with the local labor bureau for employment
Causes of unemployment	a. Economic restructuring from manufacturing to servicing industries (Made *by* Hong Kong vs. Made *in* Hong Kong) b. Asian financial crisis	a. Economic restructuring from manufacturing to servicing industries b. Macro structural change from command to market economy
The unemployed	a. Youth (aged 15–19) b. Aged (above 40) c. Less information technology–oriented	a. Aged, less-educated, single-skilled b. Problem of "hidden unemployment"—workers, often in the state sector, who have been laid off (*xiagang*)—"workers who have left their posts and are not engaged in other types of work in the same unit, but still maintain a labour relationship with the unit that they have worked" (State Statistical Bureau)

in periods of a "high tide" in employment pressures (such as the late 1970s and early 1980s), state labor bureaus have intensified problems of overstaffing in the state sector by forcing surplus urban labor power onto enterprises beyond their requirements. "Five people do the job of three" (*wu ren gan san ren gong*) was a popular saying during this period. This high degree of administrative intervention can be attributed to first, the ideological commitment to full employment, an implicit symbiosis on which the state would find it difficult to renege (White, 1987, p.116); second the tendency for enterprises to generate "excess demand" (Kornai, 1980, chapter 1), seeking to enhance their ability to meet plan targets by building up hidden labor reserves and "hoarding" labor; and the fact that it was extremely difficult to dismiss employees. The government told the workers that they were "masters" (*zhuren*) in socialist China (Kaple, 1994, p. 73). Being "the master of the firm" meant that they could not be fired no matter how poorly they had performed. Finally, the "stable" and "safe" employment system created enhanced an inherent tendency toward nepotism and the practice of "occupational inheritance" (*dingti*), that is, when a worker retired, he or she could recommend a close relative for his or her job (Howe, 1992, p. ix). Consequently, the unified employment and assignment system had lowered the quality of the employees as a whole and was implemented at the expense of lowering labor productivity. Managers' power to dismiss workers and workers' motivation to move were both weak. In terms of job mobility, the consequences in urban China after 1960 were "low levels of inter-firm transfers, high levels of regional and enterprise autarky, and risk averse strategies of advancement that discouraged firm switching" (Davis, 1992, p. 1084). This was the oft-decried "iron rice-bowl."

Emergence of New Labor Market in Shanghai

With the opening up of the economy in Deng Xiaoping's era, the Third Plenum of the Chinese Communist Party's Central Committee in 1978 introduced a comprehensive program of sweeping changes to the previous system of economic planning and management in the urban-industrial sector. Central planning and control over resource allocation, pricing, and distribution have been drawn back to permit the operation of market forces. Forms of nonstate and foreign-funded enterprises have been allowed to develop, and these have injected a powerful new competitive force into the economy.

One important impact of the reform is a general broadening of work options. State-owned enterprise (SOE), although still a major employer, is no longer the only choice available. There are collective and private enterprises, and increasing numbers are pursuing the entrepreneurial track with government encouragement (the existence of private enterprises was legitimized with the amendment to the constitution in 1988). Employment opportunities exist in Sino-foreign joint ventures or wholly owned foreign companies. As a result, the employment system became relatively more flexible. Outside the state sector, the job status of workers and the flow of labor are much closer to the notion of the labor market.

On the labor market supply side, workers have considerable freedom over their choice of employment. In contrast to the inflexible monolithic state allocation system that dominated the pre-reform era, workers enjoy multiple avenues in search of jobs, including self-initiated calls, street posters, word of mouth, and personal recommendations from friends and relatives. On the demand side, employers adopt market- and profit-driven policies. With costs and the imperatives of the labor market in mind, employers utilize various modes of recruitment and reward systems to secure workers in short supply (such as technical workers) as opposed to workers in plentiful supply (such as semiskilled and unskilled workers). To compete for skilled workers that are short of supply, enterprises would offer attractive wages. Wages and working conditions seem to direct the movement of labor from low- to high-waged jobs. Labor exchanges have been set up and private firms are able to recruit freely. In 1992, nearly one million technical and managerial personnel registered at personnel exchange centers in order to seek relocation (Westwood and Leung, 1996, p. 88). Many qualified young graduates would like to work in joint ventures. According to the statistics of the Beijing Talent Exchange Center, 70 percent out of the 4,000 technical professionals who could seek successful job transfers were from SOEs to non-SOEs (p. 88). Added to the problem of brain drain is a change in the government's policy on state-owned enterprises.

In September 1997, the Chinese government decided to transform most of its large- and medium-sized SOEs into profitable concerns within a three-year period and to change the rules whereby they were managed. The measures adopted to help SOEs overcome their critical condition involved supporting mergers, allowing bankruptcies to take place, allowing the employees to be dismissed and guided into new jobs, the introduction of competition through job losses and the pursuit of profit, and, finally, an improvement of reemployment programs all in pursuit of a competitive system in which only the best enterprises would survive (International Labour Organization, 2000). Given a legacy of overemployment, redundancies arising from enterprise reorganization with an aim of achieving profitability spread across the country. Moreover, the need for enterprises to improve and diversify their output in the face of market competition leads to a need for a highly skilled workforce. This challenge to the management of enterprises threatens a large portion of the country's labor force on account of their inadequate skills. Their skills (learned in industry) tend to be low-level, outdated, and overspecialized. Most of them are women and older workers who could offer little in the way of comparative advantages (skills) in the new labor market.

In light of the vast number of people made redundant and the inadequacy of the insurance system and social assistance available, reemployment has become a colossal task, requiring mobilization of efforts on a comprehensive scale. All state enterprises that make workers redundant are obliged to set up reemployment service centers. Loss-making enterprises can apply for state cofinancing: the Ministry of Finance and local sources such as the unemployment fund each provide one-third, and the enterprise provides the remaining third. The responsibilities of the Reemployment Centers within enterprises are as follows (International Labour Organization, 2000):

• To make a social benefit payment equivalent to the level of unemployment insurance payments to persons made redundant. After three years of unem-

Table 5.7

Layoffs in Shanghai

Nature of enterprise	Percentage (%)
State-owned enterprise	63.6
Collective enterprise	29.5
Others	6.9
Total	100.0

Causes of layoff	Percentage (%)
Enterprise financial difficulty	36.9
Enterprise re-structuring	16.7
Bankruptcy/verge of bankruptcy	8.5
Enterprise removal	8.4
Others	30.6
Total	100.0

Source: Shanghai Tongji Nianjian 1999 (Shanghai Tongji Chubanshe, 1999).

ployment, redundant staff lose their employment relationship with the enterprise and are entitled to unemployment insurance or social assistance, depending on the case.

- To pay old-age insurance.
- To pay health insurance.
- To pay unemployment insurance.
- To provide occupational information and reemployment training programs.
- To monitor the progress of redundant staff and to help them find new jobs.

The Shanghai Statistical Bureau reported that as of 30 June 1997, 260,600 laid-off workers in Shanghai needed help in employment. Table 5.7 shows the nature of enterprises from which these workers were released and the causes of the layoffs. More than 90 percent of the layoffs occurred in state-owned and collective enterprises. Around 36.9 percent of these enterprises suffered from financial difficulties, and 8.5 percent had either gone bankrupt or were on the verge of bankruptcy.

Table 5.8 shows the profile of the workers laid off in Shanghai. They share very similar characteristics to workers retrenched from factories in Hong Kong: middle-aged, low-skilled, and predominantly female. Almost half of the laid-off workers in Shanghai (47.2 percent) were age 41 and above; 72.2 percent had only junior secondary qualifications; a slight majority were female (53.4 percent); and 70.5 percent had their previous employment as workers.

Given the mismatch in labor supply and demand in Shanghai and Hong Kong,

Table 5.8

Profile of Employees Laid Off in Shanghai as of 30 June 1999

Job type	Percentage (%)
Worker	70.5
Service staff	9.1
Administrative staff	5.0
Technician	2.7
Others	12.8
Total	100.0

Education level	
Junior secondary	72.2
Others	27.8
Total	100.0

Sex	
Male	46.6
Female	53.4
Total	100.0

Age	
35 and below	20.9
36–40	31.9
41–45	29.3
45 and above	17.9
Total	100.0

Source: Shanghai Tongji Nianjian 1999 (Shanghai Tongji Chubanshe, 1999).

both cities rely on active labor market programs to cope with the problem of unemployment.

Active Labor Market Programs (ALMPs) in Hong Kong and Shanghai

Over the past forty years, "active" labor market programs have emerged as an important employment policy tool, particularly in developed countries. This policy program includes a wide range of activities intended to increase the quality of labor supply (e.g., retraining), to increase labor demand (e.g., direct job creation), or to improve the matching of workers and jobs (e.g., job search assistance). The objective of these measures is primarily economic to increase the probability that the unemployed will find jobs or that the underemployed will increase their productivity and earnings. However, more recently the case for active labor market policies has also emphasized the potential social benefits in the form of the inclusion and participation that come from productive employment (Betcherman, Dar, Luinstra, and Ogawa, 2000, p. 1). The debate around these labor market policies is often formulated in terms of the relative value of "active" versus "passive" measures in combating unemployment and its effects. So-called passive programs, such as unemployment insurance or social transfers, mitigate the financial needs of the unemployed, but they are not designed to improve their employability in any fundamental sense. On the other hand, active programs are meant to directly increase the access of unemployed workers.

Unlike Shanghai, Hong Kong does not have a policy of unemployment insurance. The Hong Kong government's primary task is "to create the conditions necessary to foster, maintain and enhance self-motivation. . . . and to assist those who have been adversely affected by the sudden economic downturn by giving them the necessary support to help themselves" (Hong Kong Government, 2000). In short, this "support" is provided by the active labor market programs, including job creation (public works, self-employment support, and wage subsidies), training, and employment services. These policies can affect labor demand, labor supply, and the functioning of the labor market in matching the two. The overall objective of these interventions is to increase employment. They can play a stabilization role in the sense of governments directly providing temporary jobs through public works or by shifting labor supply or demand curves outward by offering training or wage subsidies. Training, mobility incentives, and other employment services can reduce structural imbalances by improving the match between workers and jobs. Table 5.9 gives a concise comparison of the different active labor market programs carried out in Hong Kong and Shanghai.

Labor Market Training

This includes training where there is some form of public support. That support can come in the form of direct provision of training (e.g., through public training institutes), financial support for trainees (e.g., funding training costs and/or subsidizing trainees), or providing "infrastructure" services (e.g., labor market information, licensing, monitoring, and credential services). Governments have a range of potential roles: direct provision, regulation, providing information and standards, and financing. Many governments, like the Hong Kong government, are moving away from the role of direct provider and focusing more on addressing market failures in information and financing, while leaving more of the delivery to private providers (Betcherman, Dar, Luinstra, and Ogawa, 2000, p. 5).

Table 5.9

Comparison of Active Labor Market Programs in Hong Kong and Shanghai

	Hong Kong	Shanghai
Labor market training		
Responsible body	Employees Retraining Board (ERB)	Employees Retraining Centre
Finance	a. capital injections by government b. levy charged on companies employing imported workers	state
Design and delivery of training programs	A network of accredited training agents	state
Target of retraining	Expanded: a. focused initially on workers displaced from economic restructuring—unemployed, aged 30 and above, junior secondary education and below b. since 1993, covered housewives, the elderly, the disabled and industrial accident victims, employed persons for skill-upgrading c. since 1997, new arrivals d. since 1998, employees aged below 30	Narrow: a. *Xiagang* workers b. Unemployed
Charges	Full-time courses: free Part-time and evening courses: 40% of course fees (refundable with 80% attendance rate for low income or unemployed)	Free for *xiagang* workers and unemployed
Job creation		
A. Wage subsidy	ERB provides wage subsidy (one-third of trainee's monthly wages) to employers of tailor-made training courses:	Not applicable

                             Unskilled   Semi-skilled/skilled

(over HK$7,000 p.m.)		
Aged 30–39	1 month	3 months
Over 39	2 months	6 months
Over 50/ Handicapped	3 months	

(*continued*)

Table 5.9 *(continued)*

	Hong Kong	Shanghai
B. Direct job creation	New initiative: Government-created vacancies	Changed format: instead of direct state allocation, state acts as an intermediary
C. Micro-enterprise development assistance	Underdeveloped: a. ERB organizes courses on self-employment b. Small business loans under negotiation	Comprehensive: a. state repartitioned and renovated offices b. concessions in rentals, fees, and profit tax
Employment service		
History computerization	Long (over 30 years)	Short
a. interlinked net	√	√
b. interactive through the web	√	√
Direct contact with employers	√ (with employer's consent)	√
Service charge	X	√ (for each application form)
Private-public mix	√ (state plays a monitoring role)	X (virtually monopolized by the state)
Unemployment insurance	X	√ (maximum of 400 yuan per month)

Hong Kong

In Hong Kong, the Employees Retraining Board (ERB) is an independent statutory body set up under the Employees Retraining Ordinance in 1992. The ERB functions to provide retraining to eligible workers to assist them in taking on new or enhanced skills so that they can adjust to changes in the economic environment. ERB consists of a governing body made up of representatives from employers, employees, and the government, as well as training institutions and manpower practitioners. The Hong Kong government provided a capital injection of HK$300 million when the fund was first set up in late 1992. Three more injections of HK$300 million, HK$500 million, and HK$500 million by the government were provided in 1996, 1997, and 1999, respectively, to further expand the scheme to cope with the increasing demand for retraining (Employees Retraining Board, 2000). The fund also draws its recurrent income from a levy charged on companies employing imported workers under the Labor Importation Schemes. The levy is currently set at a rate of HK$400 per imported worker per month. Starting in 2001, an annual recurrent subvention of HK$400 million was to be allocated to the ERB from the government so that it had a more stable source of funding and could therefore draw up plans for the longer term (Hong Kong Government, 2000).

From its establishment until the end of March 2000, the ERB provided a total of 369,490 retraining places. In 1999/2000, ERB provided 78,086 retraining places. The retraining programs are funded by the Employees Retraining Board but conducted by an expanding network of accredited training bodies experienced in vocational training or in adult education such as Caritas, Young Women Christian Association (YWCA), Hong Kong College of Technology, the Hong Kong Federation of Trade Unions, and Christian Action. As of 31 March 1999, there were 52 active training bodies offering 163 retraining courses at 135 retraining centers throughout the territory (Employees Retraining Board, 2000, p. 15). A great variety of full-time, part-time, and evening courses are being offered. They broadly fall into the following categories: core course on job search skills, job-specific skills, general skills, language and computer courses, and tailor-made courses for employers. The ERB places special emphasis on tailor-made retraining courses to meet market needs. A total of forty-eight courses were developed in 1998–1999, including courses for building attendants and security guards, tunnel-boring machine operatives, and personal care workers. About 1,587 retrained completed their courses. Customized training programs for individual employers (who were involved in course design and in recruiting the trainees) proved to be highly successful with a placement rate as high as 83 percent (p. 18).

The retraining scheme focused initially on displaced workers who have difficulties finding alternative employment as a result of economic restructuring. Its major service target were those unemployed persons aged thirty and above with junior secondary education and below. The scheme was extended (in early 1993) to cover housewives wishing to reenter the job market. It was further extended to the elderly, the disabled, and industrial accident victims. The scheme extended its coverage to employed persons who need to upgrade themselves with new skills. In January 1997, the scheme was further extended to include new arrivals. As of 1 April 1998, employees below age thirty are also allowed to enroll for full-time training courses if they have difficulty finding jobs, though priority is still given to applicants aged thirty or above. Table 5.10 shows a profile of the retrainees.

All full-time courses are free of charge. In addition, retrainees attending full-time courses that last for one week or more can apply for a retraining allowance amounting to a maximum of HK$4,000 per month. Retrainees who attend part-time and evening programs have to pay a course fee of some 40 percent of the training costs. For low-income or unemployed persons, such a fee could be refunded provided that they have attained at least an 80 percent attendance rate. Job counseling and placement services are available to retrainees. Placement officers of training bodies will refer suitable vacancies to the retrainees in matching their experience, skills, personality, and job expectations. A recently established computer network between ERB and the Employment Services Division of the Labour Department also enhanced the efficiency of the system.

Table 5.10

Profile of Retrainees of the Employees Retraining Board in Hong Kong (2001–2002)

Sex	Number	Percentage (%)
Male	25,347	27
Female	67,125	73
Total	92,472	100

Age	Number	Percentage (%)
Below 30	2,843	3
30–39	30,357	33
40–49	41,519	45
50–55	13,157	14
Over 55	4,596	5
Total	92,472	100

Education level	Number	Percentage (%)
No schooling	102	0
Primary 1–5	3,011	3
Primary 6	11,656	13
Secondary 1–3	34,663	37
Secondary 4–5	35,480	38
Secondary 6–7	4,200	5
Tertiary education	3,360	4
Total	92,472	100

Source: Employees Retraining Board, Hong Kong.

Shanghai

In Shanghai, retraining is free of charge for job seekers certified to be *xiagang* workers and for the unemployed. However, if training is for self-development purposes, each trainee will have to pay 300 yuan for joining the course. The training programs, designed, financed, and delivered by the government, are linked to the needs of the market. Among the ten positions advertised to be in greatest demand for workers every week in the local newspapers, training programs will be offered for the first four ranks. Examples of such vacancies include cashiers (much sought-after for stores and supermarkets that have mushroomed in the face of the economic reforms), domestic helpers, cooks, electrical technicians, and computer technicians (Field interview, June 1999). The training centers are attached to the placement centers, widely known in Shanghai as *qiandian hou gongchang* (employment service provided in the shop, training in the backyard). Each class will have about forty trainees. The length of training differs from job to job: domestic helpers will be trained for three weeks, six hours a day; cooks and technicians will receive training for half a year, but for only three days a week. Upon completion of training, the trainees will be referred jobs. It is likely that a domestic helper employed for eight hours a day will get around 800 yuan per month. According to the head of the training center I visited, the success rate in job placement is often boosted from the normal rate of below 30 percent to as high as 70 percent for domestic helpers (Field interview, June 1999).

Job Creation

In general job creation programs are intended to support the creation of new jobs or the maintenance of existing ones. Three general types of programs fall under this category. First, there are subsidies to encourage employers to hire new workers or to keep employees who might otherwise have been laid off for business reasons. These can take the form of direct wage subsidies (for either the employer or worker) or social security payment offsets. These types of subsidies are always targeted to a particular category of worker or employer. The second category involves direct job creation in the public or nonprofit sector through public works or related programs. Typically, government funds used for these programs cover compensation costs to hire previously unemployed workers, usually on a temporary basis. Third, support is sometimes offered to unemployed workers to start their own enterprises. This can involve offering microfinancing for start-up or operating costs, allowing unemployment benefits to continue where claimants start their own business, offering grants, or providing business support services (Betcherman, Dar, Luinstra, and Ogawa, 2000, p. 6). Direct public employment creation attempts to alleviate unemployment by creating jobs and hiring the unemployed directly. It targets the displaced and the long-term unemployed, with a view to help them regain contact with the labor market, thereby minimizing the probability of stigmatization, skills obsolescence, and marginalization.

Hong Kong

In Hong Kong, the ERB has been providing wage subsidy to employers of tailor-made training courses since May 1993 (Lee, 1996, pp. 120–21). The level and duration of subsidy depend on the age of the trainee and the level of skills to be trained. For unskilled workers, the ERB will subsidize one-third of the trainee's monthly salary for one month (trainees aged between 30 and 39), two months (over 39 years of age), or three months (over the age of 50 or handicapped), depending on individual cases. Wage subsidy of half a month's salary was introduced for skilled and semiskilled workers (monthly salary of not less than HK$7,000) in July 1995. Wage subsidy for retrainees aged between 30 and 39 will last for three months, while retrainees aged over 39 will be subsidized for six months. A wage subsidy for skilled and semiskilled training is provided on the understanding that the retrainees, having passed the post-training assessment, will be employed full time at a monthly salary of not less than HK$7,000.

On the other hand, direct job creation is a new initiative taken by the Hong Kong government that adopts a market-led philosophy. The chief executive announced in his policy address that "[T]he Government will, without compromising the principle of a market economy, create new job opportunities in areas where more social investment is required. Some of the areas include:

- Additional staff for up to two years for the anti-smoking campaign and the promotion of healthy living, which will add $30 million to our annual expenditure;
- Recruiting extra workers to step up our efforts in urban cleansing and greening as well as refuse collection along the coast for a period of two years. This will improve environmental hygiene and help beautify our city. It will increase our annual expenditure by $94 million;
- Additional staff for two years for environmental improvement and community building in the 18 districts. This will cost $50 million each year;
- Increase the number of supporting staff in personal care, outreaching services and ward services to offer better services to patients. This will increase expenditure by $243 million per year in the first two years; and
- Enhance services for women, new arrivals, single-parent families, the elderly and the disabled. This will increase our annual expenditure by $228 million in the coming two years." (Hong Kong Government, 2000)

It was estimated that about 7,000 new jobs would be created by these measures. A further 8,000 jobs were expected to be available in 2001 in other expanded service areas. Subject to the Legislative Council's funding approval, the total number of job opportunities provided by the government will be around 15,000. This does not include the tens of thousands of jobs to be created by various government infrastructure projects, which will also help many jobless find work and earn a living.

Another category in job creation is Self-Employment Creation Measures or Micro-Enterprise Development Assistance. Technical assistance, credit, and other support can contribute to the creation and promotion of small-scale new businesses and self-employment. Private banks are often unable to conduct the comprehensive risk assessments required to offer credit to unemployed workers who want to create their own businesses. Public programs to support small-business loans can contribute to the removal of this distortion arising from credit rationing. To encourage people to start their own businesses, the Employees Retraining Board in Hong Kong has, since August 2000, been offering courses on self-employment on a trial basis to help the retrained start businesses or cooperative societies in areas such as providing cleaning or home help services (Employees Retraining Board Web site, 2000). The Education and Manpower Bureau of the Hong Kong government is also discussing with the ERB the establishment of a fund to provide shared facilities and other support services for the retrained who want to start up a business, and to offer one-off loans to those who can put forward concrete business plans (Hong Kong Government, 2000).

Shanghai

In comparison, the Shanghai government is more advanced in the area of self-employment creation. To create more employment opportunities, the Shanghai Labor Bureau invested in minor improvement programs at the district level, and was successful in creating over 20,000 vacancies (Field interview, May 1999). These vacancies concentrate on public hygiene (*baojian*), public security (*baoan*), environmental protection (*baolu*), and the maintenance of public facilities (*baoyang*).

The Baibang Community Service Company and the Baibang Industrial Center are examples of self-employment creation in Shanghai (Field interview, June 1999). The former will take care of the aged and the less-educated redundant workers, while the latter will cater to the needs of semiskilled workers. The Baibang Community Service Company was set up in 1994 to match the employment needs of workers laid off with the need for family service in the community. Officials who run the company are seconded from the Ministry of Social Welfare. They rely on the District Committees to identify *xiagang* workers and their appropriate skills. Government training institutes provide free training and accreditation of the worker's skills. Promotion of the service is through street posters; free local media such as newspapers, television, and radio broadcasts; and word of mouth. Services are developed according to the needs of the neighborhood, such as child care, elderly care, cleaning, and the preparation of meals. The services provided include women becoming part-time domestic help (daily serving of four different families for two hours each will mean a rather handsome amount of 1,000 yuan a month); men cleaning ventilation fans (30 yuan per fan); reading newspapers to the elderly (30 yuan for two hours); repairing household equipment such as the television, air conditioner, freezer, and the like. The company charges a commission ranging from 20 to 50 yuan from the workers, and 10 percent or 50 yuan per job from the family service recipient. As of June 1999, the company was successful in finding jobs for over 200,000 workers (Field interview, June 1999).

Another self-employment creation agency in Shanghai is the Baibang Industrial Center established in May 1998 (Field interview, June 1999). A state coal factory gone bankrupt was repartitioned and renovated into eighteen units leased to redundant worker-entrepreneurs. The government provides start-off loans ranging from 50,000 to 150,000 yuan. Employees working in the industrial center should either be workers laid off or retired from state enterprises (the ratio is about 70 to 30).

The factory units share overheads such as a meeting room for business liaisons, the cost of factory management for services such as ordering lunch boxes and sanitation matters, and clerical support for coordination with different government departments on payment of fees. They also benefit from concessions in rent, miscellaneous fees and taxes, and profit tax (first three years exempted, reduced by half for the next three years if 60 percent of the workers employed are *xiagang* workers). The idea has been considered a success as twenty-five people have left the factory to become self-employed (*geitihu*) and two others have left to set up their own business. Four branch industrial centers have been set up along similar lines of support for entrepreneurship (Field interview, June 1999).

In general, evaluations of active labor market programs indicate that few among the unemployed—usually less than 5 percent—typically take up opportunities for self-employment (Wilson and Adams, 1994). One explanation for this may be that individuals are generally risk-averse and, given a choice, will opt for other unemployment assistance such as public employment service.

Employment Services

Employment services serve brokerage functions, matching jobs with job seekers.

Hong Kong

Hong Kong is a vibrant free market economy. The Local Employment Service (LES) of the Labour Department has been rendering free and professional employment services to employers and job seekers for more than thirty years. Through an on-line computer network (since April 1998) of eleven job centers, the LES receives all types of job vacancies and registers job seekers of different backgrounds for placement. It basically operates on a semi-self-help mode whereby job seekers select suitable vacancies, register with the LES, and seek referral service from its staff. Job seekers can contact employers directly if employers are willing to open up their company names and telephone numbers while placing vacancy orders with the LES (since June 1998). A central Job Vacancy Processing Center (JVPC) was opened for employers in February 1999. The JVPC receives hundreds of vacancies a day, mostly by fax. Information is verified and entered into the computerized LES data bank within twenty-four hours and then made available to all job centers. The LES conducts regular analysis of the wage levels of the vacancies received and shares the information with all placement officers. Officers are therefore in a better position to advise employers when the terms offered are significantly below the market rate.

Since 1 April 1995, the LES has been providing personalized counseling and job matching services to unemployed local workers through the Job Matching Program (JMP). Job seekers joining the JMP are attended by placement officers who will identify for them suitable jobs according to their academic qualifications, job skills, work experience, and job expectations. Statistical performance of the JMP has indicated that it has been successful in helping the less competitive find jobs. Up to the end of 1999, the JMP registered 52,823 job seekers and secured 34,047 job offers, representing a success rate of 64.5 percent. A change in the profile of job seekers registered with the public employment service provides evidence to the structural shifts in the economy of Hong Kong. Of the job seekers registered under the program as of the end of 1995, about two-thirds were female (International Labour Organization, 2000). About two-thirds were of junior secondary education and below. Each of the age groups of 30–39 and 40–49 accounted for more than 40 percent of all registrants. Slightly over half of them came from the manufacturing sector. This profile of JMP registrants underwent some gradual changes in the following years. Among the 15,599 JMP registrants of 1998, the age groups of 30–39 and 40–49 together formed a total of 52 percent of the total registrants (International Labour Organization, 2000). Compared with the 86 percent of job seekers aged between 30 and 49 registered in 1995, job seekers now seeking assistance from the JMP are younger and better educated. The proportion of registrants from the manufacturing sector also dropped from 49 percent in 1995 to 22 percent in 1998, reflecting that workers affected by the economic restructuring are no longer confined to the manufacturing industries but are also found in the service industries. In 1998, slightly more than 59 percent of the JMP registrants had upper secondary or higher levels of education (compared with only 36 percent in 1995).

A series of new measures has also been launched by the Labour Department to further enhance its services. Its Interactive Employment Service (http://www.employment.labour.gov.hk), launched in March 1999, uses the latest technology to offer employment services on the Web. Through the Internet, employers can easily place vacancy orders with LES. Employers can also view the profiles of LES registrants to look for suitable candidates and request LES to arrange for referrals (Labour Department Annual Report, 2000, p. 105). The Labour Department also set up the Job Vacancy Processing Center in late 1998 to provide a one-stop service for employers to place vacancy orders with LES. Upon receipt of vacancy details, the staff of LES will input the information into the center's computer system. With the use of new technology, job seekers can access the vacancy information by visiting the Web site on Interactive Employment Service, or by making use of the self-service terminals installed at each LES and Labour Relations Service branch office. Job seekers who have already registered with the LES have the additional option of calling the Telephone Employment Service Center (TESC) for job referral service. Through conference calls, staff of the TESC can make arrangements for calling job seekers to talk to employers directly (p. 105).

Apart from relying on public employment service, job seekers in Hong Kong can also turn to the 1,108 private employment agencies for employment service. These private agencies are licensed and regulated by the Labour Department under the Employment Ordinance and Employment Agency Regulations, particularly with respect to fee charging. They typically provide labor exchange services to more favored segments of the labor force, such as the employed, skilled, and white-collar workers. These agencies work independently and do not forward the registries of unemployment to the Labor Bureau. They directly negotiate with the employers and match suitable jobs for their clients. The people seeking employment are considered unemployed unless they officially sign a contract with the employing company. In Hong Kong, job seekers also find out about employment opportunities through newspaper advertisements and employment services on the Internet. Such employment services are easily accessed and are user-friendly.

Shanghai

In Shanghai, private agencies are restricted and public employment service operates under near-monopoly conditions. There are 452 registered employment agencies in Shanghai, of which 338 are established under the Labor Administration Division that includes district and street organizations; 110 are operated by industries and social organizations such as trade unions, the Association for the Handicapped, and the Association of Women; and only four of them are run by private operations (Field interview, June 1999). According to the deputy head of a District Labor Bureau in Shanghai, privately operated employment agencies are not encouraged because, first, they are profit making (employment agencies run by the state are free of charge, and those run by the various industries and organizations are charged on a cost-recovery basis); and second, private employment agencies are not able to provide comprehensive services to the job seekers (Field interview, June 1999). Nevertheless, the job seekers are willing to pay the charges and buy the services from the private employment agencies because they are more approachable and efficient. It becomes easier for the unemployed to learn about the vacancies that exist. As there is a comparatively lighter workload at the private agencies, the staff can spend more time trying to match a suitable job to their clients. In fact, a user-friendly job Web site would be a big help to the unemployed.

Since 1994, the Regulations on Placement Service, enforced by the Labor and Social Security Bureau, set standards on employment services in Shanghai (Field interview, June 1999). The Shanghai Labor Bureau has been contemplating standardizing the layout, logo, and color schemes of the placement centers and the staff uniform. One placement center will be set up for every 100,000 to 200,000 residents. In the past, enterprises could only apply for labor through the government bureau. At present, enterprises can register their vacancies with the placement centers and select workers through open recruitment. Vacancies received from employers will be advertised through the interlinked net. All the state-established agencies are linked vertically (from cities, municipalities, districts, to street organizations) and horizontally among agencies at the same level. The net also linked nine employment agencies set up by the industries. Starting on 1 July 1997, state-operated employment agencies in Shanghai have been interconnected to pro-

Table 5.11

Employment Service Performance of Shanghai Labor Bureau

Year	Registered job seekers	Referrals	Vacancies	Employed	Jobs filled	Placement rate (%)
1997	125,970	125,980	143,810	21,893	80,467	17.4
1998	227,006	218,711	154,767	50,809	96,631	23.2

Source: Shanghai Labor Bureau.

vide real-time information on the job market to job seekers (Field interview, January 1999). Apart from displaying job vacancies, information on retraining is also available on the Internet. In this way, the general public can have ready access to updated labor market information and can be more focused on learning new skills. Table 5.11 shows the number of job seekers registered with the Shanghai Labor Bureau and the number of vacancies received, as well as its referral and placement rates.

Placement officers provide counseling to job seekers on labor market information, and reeducate those who possess the "iron rice-bowl" mentality to change. To establish rapport between the counselor and his or her client, individual counselors will follow through cases. Counters (differentiated by age, sex, health, and previous industries of the job seekers) have been set up for individual counseling. All counselors are trained and are required to pass a professional examination annually. As informed during the field interviews, overseas training in Germany has been arranged for outstanding counselors working at government employment agencies. The counselors were sent to Germany because Germany faced similar problems of unemployment after their reunification, and the German government had tried many means to resolve the problem. Furthermore, Germany provides free training to their Chinese counterparts.

According to the head of a placement center, it is the counselor's responsibility to understand their clients, arrange for their family needs (e.g., to arrange for babysitting or nursery care before referral to jobs), and endorse applications for unemployment insurance. The state closely monitors the situation of workers who have been "disassociated" from the state enterprises. The hardship cases are entitled to a subsistence allowance (*zuidi shenghui butei*) of 244 yuan per month when repeated referrals, counseling, and retraining do not work (Field interview, May 1999). Even then, applications for subsistence allowance are subject to regular review on a three-month basis. Yet, the Chinese government is extremely careful to balance between stringent procedures and social stability. It is the counselor's responsibility to spot all cases of genuine hardship and recommend state contribution to the old-age pension and medical insurance scheme. In any case, subsis-

Table 5.12

Monthly Rate of Unemployment Insurance in Shanghai, 1999

Cumulative contribution in years	Age of the unemployed	Monthly payment for first 12 months (yuan)	Monthly payment for next 12 months (yuan)
1 to less than 10	Less than 35	259	215
	35 and above		
10 to less than 15	Less than 35	296	237
	35 and above		
15 to less than 20	Less than 40	333	267
	40 and above		
20 to less than 25	Less than 45	270	296
	45 and above		
25 to less than 30	Less than 50	389	312
	50 and above		
30 and above	All	407	326

Source: Regulations on Unemployment Insurance in Shanghai, issued on 31 March 1999, Shanghai Shi Laodong he Shehui Baozhang Ju, *Shanghai Tongji Nianjian 1999* (Shanghai Tongji Chubanshe, 1999).

tence allowance is only payable for two years, after which the unemployed will only be entitled to a maximum unemployment insurance of 400 yuan per month depending on various factors such as age and years of contribution to unemployment insurance (see Table 5.12). The Shanghai experience of close integration of employment services with the other active labor market programs (retraining), as well as with the passive programs (subsistence allowance), can be beneficial to the extent that the unemployed acquire the skills and knowledge necessary to fill available job vacancies.

Concluding Observations

Despite enviable economic growth in Hong Kong and Shanghai, both cities face challenges in employment that originate from structural shifts in economic structures. Hong Kong moved from labor intensive manufacturing industries to a servicing economy in the 1980s and 1990s. Most displaced workers released from the secondary sector were able to secure employment because of the booming tertiary sector. The unemployment situation only worsened when Hong Kong was hit by the Asian financial crisis in late 1997. Unemployment rates soared to a thirty-year high of 6.3 percent in 1999.

Comparatively speaking, Shanghai has a fairly low unemployment rate of 3.5 percent as recorded by official statistics. Yet, the definition of unemployment in

China differs from the general conception suggested by the International Labour Organization. It excludes the non-urban sector of the population and limits its calculation to the openly unemployed, that is, those registered with the Labor Bureau. In fact, a fairly high percentage of unemployment is under the disguise of redundancies in state enterprises. The state sector, which was the only economic sector under the communist regime, served its social and political role by absorbing unemployment, leading to overmanning in virtually all state enterprises. With the macrostructural change from command economy to market economy as a result of the Open Door Policy in 1978, the non-state sector flourishes and the state sector can hardly compete with the new industries for markets (owing to heavy overheads of social responsibilities) and quality workforce. With the emergence of a relatively free labor market, talents have often been drawn from the state enterprises to the non-state sector because the former can afford more attractive salaries. As soon as the Chinese government announced in 1997 that all large- and medium-sized enterprises had to be turned into profitable concerns, the overwhelming problem of redundancies had to be faced. However, the government is cautious of any possible social instability and political upheavals should all the former "masters of the country" be pushed out of the enterprises. Voluntary redundancy measures fail because there is strong reluctance for the less competitive workers to become disassociated from the state enterprise and become openly unemployed, meaning those employment relationships with the enterprise will be severed. These workers are heavily reliant on their *danwei* for the various benefits in kind and for its contribution to the social security schemes. Hence, the problem of unemployment, open or hidden, remains to be solved.

Both governments in Shanghai and Hong Kong make use of active labor market programs to cope with the problem of unemployment, relying heavily on retraining, job creation, and employment service. Even though there is close resemblance in the nature of the programs, the role of the state is markedly different. Strong state control can be identified in all aspects of the active labor market programs in Shanghai (see Figure 5.3).

Employment service in Shanghai is virtually monopolized by the government. Private employment agencies have been suffocated by the state sector, and the labor market works within the confines of the state. Public employment centers are the major venue through which employers can advertise their staffing needs. On the other hand, public employment service in Hong Kong serves a relatively small proportion of the labor market. Job seekers can turn to private employment agencies and privately operated Web sites for employment service. The government serves the role of regulator (by licensing and inspection) to these privately operated businesses, particularly in relation to the charge of commission from job seekers. In fact, most job seekers are self-reliant and merely rely on vacancy information advertised by employers in newspapers, a mechanism that remains undeveloped in China. The Shanghai government is interventionist in strategies. The Baibang Community Service Company is a good example of the state adjusting its communist

Figure 5.3 **Comparing Hong Kong and Shanghai**

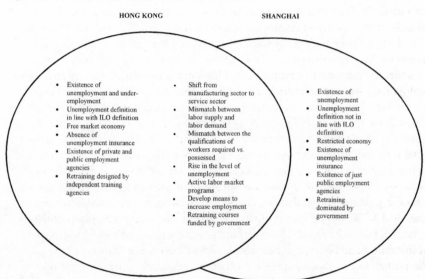

HONG KONG   SHANGHAI

- Existence of unemployment and under-employment
- Unemployment definition in line with ILO definition
- Free market economy
- Absence of unemployment insurance
- Existence of private and public employment agencies
- Retraining designed by independent training agencies

- Shift from manufacturing sector to service sector
- Mismatch between labor supply and labor demand
- Mismatch between the qualifications of workers required vs. possessed
- Rise in the level of unemployment
- Active labor market programs
- Develop means to increase employment
- Retraining courses funded by government

- Existence of unemployment
- Unemployment definition not in line with ILO definition
- Restricted economy
- Existence of unemployment insurance
- Existence of just public employment agencies
- Retraining dominated by government

political control apparatus to that of an economic intermediary. The street organizations and the district councils are used to "monitor" the situation of *xiagang* workers and identify their "marketable" skills for the provision of family services. On retraining, the government not only designs the course content and finances its operation, but also delivers the training and regulates course participation.

In Hong Kong, although retraining is heavily financed by the government in the setting up of the Employees Retraining Board (also partially financed by employer levy of HK$400 for each worker imported to work in Hong Kong), all training courses are designed and delivered by independent training agents. Employers are heavily involved in the design of tailor-made courses and the selection of retrainees. As a matter of fact, Hong Kong has long been well known for its market-led policies, meaning that the government does not seek to direct or plan the course that the economy or the markets should take, as investors and entrepreneurs are deemed to understand markets far better than officials do (Hong Kong Annual Report, 2000). The government serves only the role of facilitator and regulator. Given time and change in the philosophy of governance, the Chinese government may further relax state control and be more supportive of a free market economy.

6

Poverty and Social Security

Raymond Ngan, Ngai-ming Yip, and Wu Duo

With increasing affluence and economic growth, poverty and social assistance seem to be less dominant concerns for policy makers. This appears to be especially so in the case of Shanghai, which has witnessed a marked increase in its economic growth and wealth. In 2000, its GDP stood at U.S.$4,177 per capita, almost a quadrupling of GDP since 1978 (Shanghai Statistical Bureau, 2001). An increase of 10.2 percent in its GDP was reported in 1999 which is already the eighth year of double-digit growth (Wan, 2000). Shanghai is now commonly known as one of the major metropolitan cities in China with a plan to become China's leading financial center, as remarked by its mayor, Xu Kuangdi (Xu, 2001). Similar to Shanghai, Hong Kong experienced steady economic growth from the 1980s to 1997. Its per capita GDP was U.S.$24,073 in 2000 (Hong Kong Census and Statistics Department, 2001).

However, the Asian financial crisis, first started in Thailand in June 1997, spread rapidly to Hong Kong. It has reported that there had been a tremendous increase in unemployment cases applying to the Comprehensive Social Security Assistance Scheme (CSSA)—the only form of public measure and cash benefits available to aid the poor (Social Welfare Department, 1998). The increase was so alarming that it drew the attention of policy makers and led them to introduce an Active Employment Program so as to encourage the able-bodied CSSA recipients to return to work (Social Welfare Department, 2000). The director of Social Welfare, Carrie Lam, made the following remarks:

> In the ten-year period from 1992–93 to 2001–02, total recurrent expenditure on social welfare increased from HK$7.6 billion to HK$30 billion [U.S.$1=HK$7.82]. As a result, welfare's share of total recurrent public expenditure increased from 8.3% to 13.8%. Of the estimated expenditure of $30 billion for 2001–02, about two-thirds or $21 billion will be spent on social security schemes providing financial support to those in need. But $30 billion are a significant share of public resources. . . . There is a general consensus within the community that we do not

wish to see Hong Kong become a welfare state relying on heavy taxes, that there is a limit to how much the Government can spend and that the virtues of self-reliance, family cohesion and community support should be preserved. (Lam, 2001)

Shanghai, too, has its eyes on the amount of increased expenditure to aid the unemployed and the poor. The Shanghai Statistical Bureau reported that in June 1997, there were 260,600 laid-off workers in Shanghai needing help finding employment. Thirty-seven percent of their employing state and collective enterprises suffered from financial difficulties, and 8.5 percent were either already bankrupt or on the verge of bankruptcy (Shanghai Statistical Bureau, 1999). With the Chinese government's proclaimed policy to transform most of its large- and medium-sized state-owned enterprises (SOEs) into profitable ones within a three-year period, the chance of massive unemployment and redundancy was expected to be inevitable (Zhu, 1998). How to reform the state's hitherto huge expenditure on social security payments into a Multi-Tier Social Insurance Fund being financed by workers' own contributions, their employing enterprises, and local counties' and the state's contributions was already being considered in 1993 in Shanghai by the Shanghai Social Insurance Bureau (Field interview, June 1999).

The case of public expenditure on welfare to help the poor was revisited by both governments since 1998 with increasing concern. However, the concern was driven less by a humanitarian concern for the well-being of the poor and the unemployed than by a growing apprehension about the huge amount of public funds being spent on the dole with a view, in particular, on how to encourage the able-bodied welfare recipients to return to work as soon as possible. This concern had support from the introduction of the Active Employment Program and the "Support for Self-reliance Scheme" in Hong Kong which aim at encouraging and assisting CSSA (Comprehensive Social Security Assistance) unemployed recipients to regain employment and move toward self-reliance (Social Welfare Department, 2001). The city government of Shanghai has also been launching Active Labor Market Programs through free labor market training courses for unemployed, redundant workers and job seekers by setting up a large number of Re-employment Service Centers to promote the chances for employment and self-employment (Fudan University, 1998). We first examine the political-economic context for reforms in social security programs in the two cities—Shanghai and Hong Kong. Second, we examine reforms in the social security programs by the two city governments to help the unemployed poor. And finally, an evaluation of the success or failure of such programs in the context of rising unemployment and a political economic analysis is discussed (Phillipson and Thompson, 1996; Walker, 1981).

The Political Economy Context for Reform

Hong Kong: From Dependence on the Dole to Support for Self-Reliance

An official poverty line does not exist in Hong Kong as the SAR Government has been reluctant to carry out such a study (Ngan, 1996). The CSSA recipients, being those with the most meager rates of subsistence living, are commonly viewed as the poorest sectors in Hong Kong. The total number of CSSA cases had almost been increased four times from 59,900 household cases in 1985 to 233,000 household cases in 1999 (Hong Kong Government, 1999). The number of CSSA recipients had increased from 153,000 people or 2.5 percent of the total Hong Kong population in 1995 to 3.4 percent in 2000/01 (Hong Kong Census and Statistics Department, 2001). But what is of increasing alarm to policy makers in Hong Kong is the highest increase in CSSA expenditure from HK$1.408 billion in the 1992/93 financial year to HK$14.407 billion in 2001/02, an increase of 9.6 times in an interval of eight years. Table 6.1 shows that government's spending on CSSA had increased from 31.6 percent of the total social security expenditure in Hong Kong in 1992/93 to 72.8 percent in 2001/02 (Hong Kong Census and Statistics Department, 2002). Correspondingly, social welfare expenditure (which includes social security payments) expressed as a percentage of total public expenditure, had increased from 3.9 percent in 1992/93 to 7.25 percent in 2001/02. But CSSA payments alone constituted the largest increase in social welfare expenditure, as shown in Figure 6.1. When expressed as a percentage of total government recurrent expenditure, CSSA expenditure increased rapidly from 2.6 percent in 1993/94 to 8.6 percent in 1999/2000, but then dropped to 5.3 percent in 2001/02 (Hong Kong Census and Statistics Department, 2002).

It was in the above-mentioned context of a continual alarming increase in CSSA expenditure that the government eventually released the CSSA Review Report titled "Support for Self-reliance" for public consultation in December 1998. Besides, economic performance for the Hong Kong Special Administrative Region since 1997 was not satisfactory. Following the onset of the Asian financial crisis in July 1997, the financial secretary in his 1999–2000 Budget Report, noted:

> Overall investment receded in 1998, after achieving double-digit growth for four years in a row. Investment spending for the year dropped by 5.8% in real terms. Looking forward, the external environment could still be difficult in 1999. I expect the economy will still be slack in the early part of 1999, with a more visible pick-up in the latter part. (Hong Kong Government, 1999)

The financial secretary went on to remark:

In the past five years to 1998–99, we have increased recurrent spending in social welfare by 103%. This significant growth in expenditure on welfare over the years has increased the share of social welfare recurrent spending from 8.7% in 1994–95 to its current 14.1 in 1999–2000. We would only be able to meet the ever-growing demands within our overall expenditure constraints by re-examining policies and re-adjusting spending priorities. The modifications to the CSSA Scheme which aim at removing dependence on CSSA and encouraging able-bodied recipients to rejoin the workforce are introduced with this objective in mind. We cannot afford much longer double-digit growth in welfare spending each year. (Hong Kong Government, 1999, p. 29)

It was under this context of a rapid and alarming increase of the able-bodied unemployed people on the dole that the SAR government in Hong Kong introduced measures for reform in her social assistance programs. In December 1998, the Social Welfare Department published the Report on Review of the CSSA Scheme (Social Welfare Department, 1998). Hallmarks of prominent measures recommended for implementation were

1. To implement a "Support for Self-reliance" Scheme (SFS) to encourage and assist CSSA unemployed recipients to move toward self-reliance. First, the scheme comprises an Active Employment Assistance program, replacing the requirement for the unemployed applicants to register with the Local Employment Service of Labor Department; second, a Community Work program to offer additional help by way of community work; and third, incentives to work through the provision of enhanced disregarded earnings.
2. To strictly enforce the policy to terminate CSSA payment to an unemployed recipient who fails to comply with stipulated requirements.
3. To reduce standard rates for able-bodied adults in households composed of three or more such members.
4. To strengthen the existing arrangements to prevent fraud and abuse.
5. To require CSSA single parents with the youngest child aged twelve and above to seek work or do community work (Social Welfare Department, 2001).

The effectiveness of such measures is discussed in subsequent sections. But it has indirectly molded Hong Kong's Comprehensive Social Security Assistance Scheme into one that is increasingly marked with an overcoat of "market workfare," especially for the able-bodied unemployed welfare recipients (Grover and Stewart, 1999). It has also adopted a seemingly global trend of reforming social security with a "welfare-to-work" strategy as initiated by the New Labour Government in the United Kingdom (Grover and Stewart, 2000). By placing greater emphasis on workfare or welfare-to-work for social assistance recipients, such policy initiatives often place greater weight on the reduced number of able-bodied un-

Table 6.1

Social Security Expenditure in Hong Kong

	1992/93 $ million	%	1993/94 $ million	%	1994/95 $ million	%	1995/96 $ million	%	1996/97 $ million	%	1997/98 $ million	%	1998/99 $ million	%	1999/2000 $ million	%	2000/01 $ million	%	2001/02 $ million
Comprehensive social security assistance	1,408.5	31.6	2,443.4	43.1	3,426.8	50.8	4,831.1	55.9	7,127.8	63	9,441.3	67.4	13,028.7	72.6	13,623.4	73	13,559.8	71	14,404.6
Disability allowance	830.4	18.6	780	13.8	773.4	11.5	914.7	10.6	1,036.6	9.2	1,181.8	8.4	1,321.0	7.4	1,419.8	7.6	1,567.1	8.2	1,659.5
Old age allowance	2,141.6	48	2,337.1	41.3	2,435.6	36.1	2,768.0	32	3,005.0	26.6	3,238.4	23.1	3,416.3	19	3,463.5	18.6	3,562.5	18.7	3,581.2
Compensation schemes	82.9	1.9	103.3	1.8	111.5	1.7	124.1	1.4	136.4	1.2	153	1.1	183.2	1	146.7	0.8	410.9	2.1	152.7
Total social welfare expenditure	4,463.4	100	5,663.8	100	6,747.3	100	8,637.9	100	11,305.8	100	14,014.5	100	17,949.2	100	18,653.4	100	19,100.3	100	19,798
Total public expenditure ($ million)	113,332		147,438		164,155		183,158		182,680		194,360		239,356		214,500		278,388		273,151
of which % social welfare expenditure	3.90%		3.8%		4.10%		4.70%		6.20%		7.20%		7.50%		8.70%		6.86%		7.25%

Source: HKSAR, *Hong Kong Annual Digest of Statistics*, 1998, 2000, 2001, and 2002. www.info.gov.hk/swd/html_eng/index.html.

Note: Compensation schemes include (1) Criminal and Law Enforcement Injuries Compensation; (2) Traffic Accident Victims Assistance; and (3) Emergency Relief.

employed people on the dole, tend to neglect the plight of the needy and the poorest, and also neglect to acknowledge structural factors leading to massive unemployment. Grover and Stewart (2000) commented:

> This modernization process in social security has not involved questioning the ways in which the system might be used to overcome some of the worst aspects of capital accumulation—unemployment, "flexible" employment and low wages— but has involved a process of questioning how in the context of these phenomena the commitment of economically inactive people to such a labor market can be maintained.

Underneath the rising expenditure on the dole, an analysis of labor force statistics shows that there was not only a sharp increase in the number of unemployed people in Hong Kong but also a spread to service industries and a higher ladder of occupations. According to labor force surveys, the unemployment rate has risen from 2.2 percent or 69,000 people in July–September 1997 to 7.4 percent or 253,000 people in March–May 2002, an increase of 3.3 times in less than five years (Hong Kong Census and Statistics Department, 2002). In the same period, the underemployment rate has risen from 1 percent or 33,100 people to 3.1 percent or 108,000 people, respectively. Not only was there a sharp increase in the number of unemployed but the widespread effects of the unemployment problems were revealed in the following ways: First, unlike the unemployment problem in the 1980s when the most affected people were in the manufacturing industry and other low-level occupations, the current unemployment problem affects the service industry employing the largest working population in Hong Kong and also people higher on the ladder of occupation (Li, 1999). Second, more people were already unemployed for over a year; in fact, over half of those unemployed people above the age of forty were reported being so (Hong Kong Council of Social Service, 2000). Third, the unemployment rate for those people aged forty and above had increased three times from 25,900 persons in the last quarter of 1996 to 66,900 persons in the last quarter of 2000 (Hong Kong Census and Statistics Department, 2001).

The seriousness of the unemployment problem especially after the Asian financial crisis is beyond dispute. Nevertheless, the financial secretary, Donald Tsang (1998), remarked that "the answer to the unemployment problem, both here in Hong Kong and elsewhere, lies with economic recovery. We must not undermine our business environment by overturning our fiscal prudence or increasing the risks of costs of doing business in Hong Kong." The fiscal constraint is restraining a fundamental examination of factors leading to structural unemployment in Hong Kong to help reduce the number of able-bodied unemployed on the dole.

Figure 6.1 **CSSA Expenditure as a Percentage of Total Government Recurrent Expenditure, 1993/94–2001/02**

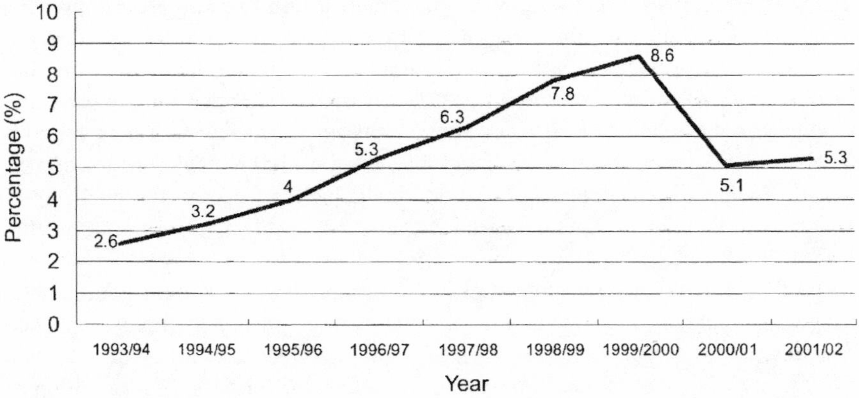

Source: Census and Statistics Department, HKSAR (various years), *Hong Kong Annual Digest of Statistics.*

Note: Social Welfare Expenditure includes (1) Comprehensive Social Security Assistance (CSSA), (2) Disability Allowance, (3) Old Age Allowance, and (4) Compensation Schemes.

Shanghai: From Unitary Financing to Multitier Financing in Social Security Payments

A System of Three Basic Minimum Subsistence Levels

Unlike Hong Kong, Shanghai has a well-defined official poverty line. A system of three basic minimum subsistence levels has been established since the 1990s (*Shanghai Encyclopaedia*, 1999; Qiao, 1999). As a first and basic protection, Shanghai was the first urban city to have introduced the "minimum-living-standard guarantee scheme" (*zuidi shenghuo baozhang*) in 1993. This is a first and basic subsistence living protection to those citizens living in urban counties with or without an income below the official poverty line. As of December 1998, 536 cities had established this subsistence living protection system, which was 80 percent of the total number of cities in China. A total of 2.35 million people have received such assistance throughout China (*China Labor and Social Security News*, December 1, 1998; Mok and Huen, 1999). Being a basic subsistence living allowance, the cash benefit was set at a minimum level. In Shanghai, it was set at 185 yuan per month in 1996, 195 yuan in 1997, and at 280 yuan from July 1999 onward (Field notes and interview with Shanghai Ministry of Civil Affairs, 2000; *Shanghai Encyclopaedia*, 1999).

Unemployment insurance is the second line of income protection for unemployed and laid-off workers. It was introduced in Shanghai in 1986, with a level of assistance set at 75 percent of the minimum wage for those laid-off workers for the first twelve months, and 60 percent for the thirteenth to twenty-fourth months. In 1997, it was 236 yuan for the former and 195 yuan for the latter entitlement. Since June 1998, redundant workers (*xiagang zhigong*) would be entitled to a basic living allowance for a maximum of three years provided that they register and participate in reemployment service centers, with an assistance standard a little bit higher than the unemployment insurance but not too high to affect their work incentives. Generally speaking, the amount should not fall below the level of unemployment insurance (Saunders and Shang, 2001; *Shanghai Encyclopaedia*, 1999).

The third line of income protection is the minimum wage system. Since 1993, Shanghai has already implemented the minimum wage for urban employment. The level was 270 yuan per month in 1995 in Shanghai, 300 yuan in 1996, 315 yuan in 1997, and 325 yuan in 1998 (*Shanghai Encyclopaedia*, 1999). The setting up of the minimum wage is a statutory requirement for urban workers in China whereas the SAR Government in Hong Kong still declines this requirement, worrying that it would upset the economy with higher production costs and that the minimum wage would most commonly become the optimal and modal wage for workers.

At first glance, it appears that this system of three lines of basic income guarantees in Shanghai seems to be a more well-structured and well-defined social assistance system than Hong Kong's CSSA system. However, the level of assistance is very low and barely sufficient for a minimum subsistence living, especially for the first minimum-living-guarantee scheme. According to the World Bank (1996), China's poverty line, expressed in U.S. dollars (U.S.$1.00 = 8.2 yuan) and using exchange rates designed to reflect "purchasing power parity" (PPP), was as low as U.S.$.60 per person per day in 1992. The World Bank commented that the $.60-per-day poverty line used in China is one of the most stringent in the world. In much of Latin America, the line is set at $2.00 per day. Expressed in Chinese currency, the $.60 poverty line amounted to 147.6 yuan in 1992. Thus, 185 yuan in 1996, as promised by the first minimum-living-guarantee scheme in Shanghai, appeared very close to basic poverty subsistence relief. It only allowed for a living of basic foodstuffs and no other living styles. Second, eligibility criteria for unemployment insurance excludes the voluntary unemployed, that is, those workers who quit a job on their own and are awaiting reemployment. Leung (1996) commented that the beneficiaries of the unemployment insurance program only include employees who are laid off by enterprises or who lose their jobs due to termination of contracts or bankruptcy.

From Cash Relief to Reemployment Service Centers

Although the government in Shanghai is apprehensive of the increasing amount of public funds spent on the dole, it was the magnitude of the problem of unemployment and the need to speed up the reemployment of *xiagang* personnel (redundant and laid-off workers) in state and collective enterprises that drew the attention of the municipal government in Shanghai to introduce a series of social security reforms and reemployment service centers since the mid-1990s. Officially, *xiagang* personnel refers to "workers who, because of the situation relating to production and operation of their employing units, have left their original work posts and have not engaged in any other work in their work units, but still maintain labor relation with their original employers" (Guangdong Social Insurance Provincial Bureau, 1999). Table 6.2 shows that the urban unemployment rate had increased sharply from 1.5 percent or 77,000 persons in 1990 to 2.9 percent or 159,600 persons in 1998 (Shanghai Statistical Bureau, 1999). However, this figure does not include the number of redundant workers (*xiagang zhigong*). As remarked by the mayor of Shanghai, there were already 860,000 redundant workers during the years 1992–1996. Wu (2000) estimated there were already about 1.9 million redundant workers in Shanghai in the third quarter of 1999 (*Wen Hui Daily News*, 26 November 1999). The problem of introducing relief measures for them and helping them get reemployment was a thorny concern for the municipal government (Fudan University, 1998). In 1996, there were already 698,900 people in Shanghai receiving unemployment insurance, with a total spending of 83.67 million yuan (Shanghai Statistical Bureau, 1999; and Table 6.4). As a result, reemployment service centers were first set up in Shanghai in 1996 on a large scale in order to widen channels of training, reemployment, and self-employment opportunities. As discussed by the mayor of Shanghai, these centers were able to bring 660,000 laid-off workers back to employment (Fudan University, 1998, p. 319). In 1997, the State Council in China cited the successful experience of Shanghai in implementing the massive reemployment programs as a model for other cities. These centers look after redundant and laid-off workers for a period of not more than three years during which a basic living allowance would be paid to them. The centers would also pay their medical and social insurance fees, but these workers have to attend job training and retraining courses and attend to jobs introduced to them. The mayor of Shanghai was proud to share the city's success and pioneering role in developing these Reemployment Service Centers (Xu, 2001).

The central government's strong determination to promote reemployment among redundant and laid-off workers in bankrupting state and collective enterprises is reflected by the massive and widespread setup of 124,000 reemployment service centers in SOEs (state-owned enterprises) throughout urban areas in China in the third quarter of 1998, an increase of 104,000 from the second quarter. The number of laid-off workers cared for in these centers also increased by 5.45 million to a total of 7.01 million people, with the total amount of living allowance given at 3.1 billion yuan (Mok and Huen, 1999; Qiao, 1999).

Table 6.2

The Number of Unemployed People in Urban Areas and the Unemployment Rate in Shanghai (1990–2000) (Expressed in 10,000 People)

Year	Employment in urban areas	Unemployed in urban areas	Unemployment rate (%)
1990	519.44	7.70	1.5
1991	532.32	7.61	1.4
1992	536.24	9.42	1.7
1993	521.28	12.97	2.4
1994	521.06	14.85	2.8
1995	531.34	14.36	2.7
1996	533.00	14.54	2.7
1997	534.91	14.90	2.8
1998	566.99	15.96	2.9
1999	540.92	17.47	3.1
2000	580.68	20.08	3.5

Source: Shanghai Municipal Statistics Bureau, *Shanghai Statistical Yearbook* (China Statistics Press, 2001).

From Noncontributory Social Insurance to a Multitiered Financing Old Age Pension System

Table 6.3 shows that in Shanghai, labor insurance and welfare funds for workers rose from 46.30 percent of total wages for workers in 1990 to 50.27 percent in 1998 (Shanghai Statistical Bureau, 1999, p. 47). The increase was particularly higher for state-owned and collective enterprises, constituting up to 56.9 percent and 71.4 percent of total payrolls for workers in 1998 (Table 6.4). This is much higher than the national average of 30.4 percent in 1997 (China Statistical Bureau, 1998; Ngan, 1999). Expressed in money terms, labor insurance and welfare funds had increased markedly in Shanghai from 6.8 billion yuan in 1990 to 25.657 billion yuan in 1998, an increase of 3.8 times. It is not only the magnitude of increase but also the vast sum of expenses being paid for such funds that drew the government's attention to introduce fundamental reforms in the Old Age Pension Scheme since 1993. Chow (2000) observed that in the first half of the 1980s, expenditure on China's Labor Insurance program had been increasing at a much faster rate than the increase in total wages. The expenditure on labor insurance and welfare amounted to 7.81 billion yuan in 1978 for the country as a whole, representing 13.7 percent of the total wage bill, and increased to 33.16 billion yuan in 1985, or 24 percent of the total wage bill; it rose again in 1995 to 236 billion yuan or 29.2 percent of the total wage bill.

The reasons why Shanghai spent more payments in pension and welfare expen-

Table 6.3

Labor Insurance and Welfare Funds for Staff, Shanghai (1990–2000) (Expressed in 100 Million Yuan)

Indicators	1990	1996	1997	1998	1999	2000
Total	68.00	245.50	252.61	256.57	287.70	347.86
Percentage of total wages	46.30	49.80	49.50	50.27	49.3	56.6
Labor insurance and welfare funds for non-working staff and workers	37.22	161.83	168.35	174.17	215.46	250.93
Pensions for retired veteran cadres	0.97	7.01	5.02	5.60	7.17	7.32
Pensions for retirement	18.25	99.05	113.55	126.62	166.62	195.90
Living expenses for the resigned	0.25	0.94	1.14	1.12	1.74	1.92
Medical care	7.42	40.39	34.67	28.18	31.01	34.39
Others	10.33	14.44	13.97	12.65	8.92	11.40
Labor insurance and welfare funds for working staff and workers	30.78	83.67	84.26	82.40	72.24	96.93
Subsidies for collective welfare funds	5.49	15.12	15.11	12.74	7.74	8.54
Expenses for cultural activities, sports, and propaganda	0.44	1.26	1.43	1.25	1.13	1.34
Medical care	12.98	42.91	46.15	46.29	40.11	66.66
Others	11.87	24.38	21.57	22.12	23.26	20.39
Annual average per capita (yuan)	612	1,811	1,887	1,947	1,751	2,432

Source: Shanghai Municipal Statistics Bureau, *Shanghai Statistical Yearbook* (China Statistics Press, 2001).

diture than the country as a whole are first that Shanghai is China's metropolitan city with the fastest growth in an aging population. Gui (2000) observed that those aged sixty-five and above were already constituting 13.8 percent of the total population in Shanghai. By 2030, this would increase to a range of 26.6 percent to 30.4 percent (see Figure 6.2). Second, a gradual increase of pension level for retired workers (see Table 6.5) shows that the average monthly pension in Shanghai had increased from 267 yuan in 1993 to 554 yuan in 1998, a slightly more than twofold increase in five years (Wu, 2000).

Following the State Council's decision in 1991, a new Old Age Pension System was to be implemented to revamp the non-contributory Labor Insurance system that had been enforced in China since 1953 (*Beijing Review*, 2000). A Multitiered Financing Old Age Pension system was to be introduced with contribution from workers themselves and from their employing enterprises. In the past, the Social Insurance Fund was totally financed by collecting contribution from

Table 6.4

Labor Insurance and Welfare Funds for Staff and Workers Classified According to Types of Enterprises, Shanghai (2000)

Indicators	Labor insurance and welfare funds	State-owned units	Collective-owned units	Other units
Total	347.86	252.16	46.59	49.11
Percentage of total wages	56.6	67.9	118.4	24.1
Labor insurance and welfare funds for non-working staff and workers	250.93	188.20	39.64	23.09
Pensions for retired veteran cadres	7.32	6.76	0.17	0.39
Pensions for retirement	195.90	144.42	35.24	16.24
Living expenses for the resigned	1.92	0.72	1.07	0.13
Medical care	34.39	27.72	2.43	4.24
Others	11.40	8.58	0.73	2.09
Labor insurance and welfare funds for working staff and workers	96.93	63.96	6.95	26.02
Subsidies for collective welfare funds	8.54	5.05	0.56	2.93
Expenses for cultural activities, sports, and propaganda	1.34	0.96	0.05	0.33
Medical care	66.66	44.29	5.32	17.05
Others	20.39	13.66	1.02	5.71
Annual average per capita (yuan)	2,432	2,710	1,419	2,291

Source: Shanghai Municipal Statistics Bureau, *Shanghai Statistical Yearbook* (China Statistics Press, 2001).

collective and state-owned enterprises. Workers now have to contribute 3 percent of their monthly wages to the Old Age Pension Fund, plus a 25 percent contribution from their SOEs. The contribution would be forming collective retirement funds in provinces, to be administrated by Provincial Social Insurance Bureaus, and forming a new old age protection system comprised of three composite accounts for each participating worker: the basic pension fund, the enterprise sup-

Figure 6.2 **Trend and Projection of Aging Population in Shanghai, 1979–2030**

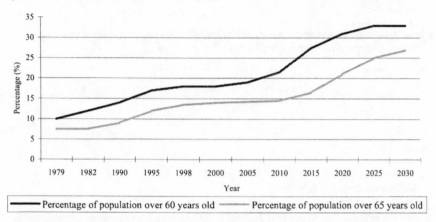

Year

| Percentage of population over 60 years old | Percentage of population over 65 years old |

Source: Historical figures are provided by China Statistical Press, *Collection of Data on the Elderly Population in Shanghai*; 1998 statistic is provided by the Shanghai Municipal Public Security Bureau; projected figures are provided by Shanghai Research Center on Aging.

plementary pension fund, and the workers' individual pension-savings fund to replace the enterprise-based system under the old Social Insurance scheme (Tang and Ngan, 2001).

The new Old Age Pension system was started in 1993 in Shanghai with the setting up of the individual pension account for participating workers. Each worker was required to contribute 2 percent of his/her monthly wages to the new pension fund. The government had the intention of gradually increasing the worker's contribution by 1 percent every two years, so that eventually the contribution load by collective enterprises would be lowered to about 10 percent to 15 percent from their existing level of 25 percent to 30 percent of workers' payroll (Field interview with the Shanghai Social Insurance Bureau, 2000). In 1998, the workers' contribution rate was 5 percent of their wages (*Shanghai Encyclopaedia*, 1999). The pace of increase was slightly slower than was initially anticipated, and the corresponding reply from officials was that the lowering of economic growth for the city had to be taken into consideration.

The mayor of Shanghai remarked that pension reform is one of the main directives in Shanghai's Five Year Development Plan and is serving a stabilization function in the city's restructuring process (Xu, 2001). The setting up of individual pension accounts in the new Multitiered Financing Old Age Pension Funds helps collective enterprises reduce their heavy commitment in social security expenditure (*Shanghai People's Daily*, 17 March 1997).

Table 6.5

Amount of Monthly Pension among Urban Retirees, Shanghai (1993–1998)

Year	1993	1994	1995	1996	1997	1998
Monthly pension amount	267.3	342.9	405.4	462.0	531.1	554.5

Source: 1999 Shanghai Care for the Aged Seminar Paper by Ministry of Civil Affairs and Wu (2000).

From State Financing to Multitier Financing in Unemployment Insurance and Welfare Relief

In 1997, CCP secretary-general Jiang Zemin announced the ushering in reforms of state-owned enterprises to make them more competitive (Jiang, 1997). This would involve privatization of those bankrupting SOEs, and calls for eventual self-financing of other SOEs. Jiang anticipated that this restructuring process would inevitably result in a main rescheduling of SOEs and collective enterprise workers, with some being laid off or redundant. It was anticipated that the bills for unemployment insurance would be increasing, with more people getting laid off. The Ministry of Labor and Social Security estimated that in 1998, for the country as a whole, more than 11.5 million workers were idle and another 3.5 million would also lose their jobs (Chow, 2000).

Unemployment insurance was set up in Shanghai in 1986. It was financed entirely by a contribution of 0.6 percent of the total payroll from participating state enterprises providing cash assistance to workers in times of involuntary unemployment—those who are laid off by collective enterprises or lose their jobs due to the termination or cancellation of contracts, as well as the bankruptcy or near-bankruptcy of enterprises. Workers were not required to contribute and local governments are supposed to cover any deficit that the funds incur (Leung, 1996). The national policy is that an unemployed worker with more than five years of continuous working experience can get 60 percent to 75 percent of his standard wage for a year, and then 50 percent for another year, with a maximum period of entitlement of 24 months. Those with one to five years of continuous working experience can obtain only one year of assistance. Unemployment insurance in Shanghai was provided at a higher level—75 percent for the first year and 60 percent for the second year (*Shanghai Encyclopaedia*, 1999).

With more bankrupting SOEs' workers getting laid off, the drain on this noncontributory unemployment insurance had alerted the attention of policy makers. The contribution rate from participating SOEs was raised from 0.6 percent to 1 percent in 1993 with the intent to include all urban employees, not just SOE workers. In 1993, eligibility for unemployment benefits was extended to include merger

Table 6.6

Amount and People under Welfare Relief, Shanghai (1996–1998)

Year	No. of people (10,000 persons)	Amount (in 10,000 yuan)			
		Minimum-living standard guarantee	Relief in kind	Voucher of free oil and basic foodstuff	Total
1996	50.1	654	2,703	10,345	14,933
1997	48.2	1,716	3,496	14,257	19,470
1998	46	2,654	2,880	13,800	23,834

Source: Ministry of Civil Affairs, Annual Report.

or restructuring SOEs. This had increased the number of eligible unemployed workers from 100,000 in 1990 to 1.27 million in 1995 (Chow, 2000; Wang, 1998). As a result, the minister for labor announced in 1997 that it would be necessary to raise the contribution rate of the enterprises for unemployment insurance from the present 1 percent to 2 percent, and workers would also need to make their contributions, at 1 percent of their salary (Ministry of Labor, 1997). Thus the hitherto noncontributory formulae were changed to a multitiered financing one with contribution from enterprises, workers, and local county and provincial governments when necessary. According to the *Shanghai Encyclopaedia* (1999), there were 17,989 enterprises participating in the unemployment insurance, covering about 4 million workers in Shanghai in 1998. Total expenses for labor insurance and welfare funds for working staff and workers, notably unemployment living expenses for laid-off and redundant workers, had increased markedly from 30.78 million yuan in 1990 to 96.93 million yuan in 2000, as shown in Table 6.4 (Shanghai Statistical Bureau, 1999). Whether the new funding formulae have some bearing in affecting this remains to be seen. But the successful experience in Shanghai with the setting up of massive Reemployment Service Centers in promoting training, cash assistance, and job-seeking opportunities for redundant and laid-off workers does appear to have a close relationship with the moderation of expenses on unemployment insurance. The mayor of Shanghai was proud to share the city's success in getting 660,000 workers back to work in 1992–1996 when compared with 860,000 redundant and laid-off workers then (Xu, 2001).

Table 6.6 shows the number of people under welfare relief, in cash and in kind, in Shanghai from 1996 to 1998 (Ministry of Civil Affairs, 1999). The total number of people under welfare relief appears not too high, with 460,000 to 500,000 destitute people, but the amount of public spending looks large, from 149 million yuan in 1996 to 238 million yuan in 1998, an increase of 1.9 times in only two years. Thus the Shanghai government has called forth the need for multiple forms of financing.

Welfare relief mainly covers the nonworking poor, destitute, homeless, and childless elderly people, as well as the disabled and the chronically ill. It also includes assistance paid to those poor people in the form of a "minimum-living-standard guarantee" system. The central government of China declared in the 1990s that welfare relief should be borne mainly by provincial and local county governments. The Shanghai government went one step further by proclaiming in 1995 that material relief belongs to the responsibility of SOE and collective enterprises and local street relief offices where destitute people were found living. However, two major problems had hindered this move. First, such a policy added up the operating cost of most enterprises, and, second, those destitute living in already bankrupting and near-bankrupting enterprises usually ended up getting no assistance. As a result, since 1 July 1999, the minimum-living-standard guarantee expenses would be borne by the metropolitan and county governments of Shanghai (each shouldering 50 percent). This has now turned out to be an effective and reliable relief system as remarked by the deputy director of civil affairs in Shanghai (Field interview, 2000). However, the Shanghai government also encourages local non-profit-making organizations to give a helping hand in providing material assistance to the destitute and the needy at times of natural disasters. Other local initiatives were also encouraged to participate actively in this endeavor, including appeals to the Shanghai trade unions, the Shanghai Federation of Women, the Shanghai Charity Fund Organization, Shanghai Fund for the Aged, Shanghai Welfare Relief Fund Organization, and the Shanghai Community Chest in order for welfare relief to also be funded by local and county efforts.

This analysis shows that recent increasing concern about cash assistance for the poor was motivated more by the alarming increase in public expenditure on the dole as in Hong Kong's CSSA payments, and the massive number of people becoming unemployed, plus a mounting pension payment higher than the national average as in the case of Shanghai. Fundamental reforms were introduced in social assistance schemes, unemployment insurance, and pension funds in the 1990s. However, there was little inquiry on the adequacy of the levels of cash benefits to help the destitute, but more concern on how to prevent fraud and abuse in Hong Kong and on how to get redundant and laid-off workers back to work quickly. The success of such reforms is described in the paragraphs that follow.

How Effective Is the "Support for Self-Reliance Scheme" in Hong Kong?

A major reform in Hong Kong's cash assistance scheme for the poor is the introduction of the Support for Self-Reliance (SFS) Scheme in June 1999, with the aim to encourage and help unemployable CSSA recipients find work and become self-reliant. The SFS Scheme comprises three components:

1. Active Employment Assistance Program (AEA). Caseworkers in Social Security Field Units will provide active employment assistance to able-bodied, unemployed CSSA participants to develop personal plans to find jobs, provide them with information on employment services and retraining courses available, and refer them to appropriate services.
2. Community Work Program (CW). CSSA unemployed recipients are required to perform unpaid community work for one full day (8 hours) or two half-days (4 hours x 2) a week. Community work includes either community services such as simple library/clerical work or environment-related projects such as cleaning country parks and beaches.
3. Enhanced Disregarded Earnings (DE). The first month's earnings from a full-time job would be disregarded totally during a two-year period in addition to the provision of monthly disregarded earnings. This means that when a CSSA recipient gets a job, the first month's salary would still not affect his/her entitlement to CSSA benefits.

These new initiatives appeared to be working well as confirmed by an evaluation report prepared by the Social Welfare Department. It was presented for discussion at the Legislative Council on 11 July 2001. Major evaluation findings were cited as below (Social Welfare Department, 2001):

1. The unemployment CSSA caseload, which is the target of the SFS Scheme, recorded month-to-month decrease ranging from 1 percent to 3 percent during the 12-month period since the implementation of the Scheme in June 1999.
2. There were 24,998 unemployment cases in May 2000. This represents a drop of 23 percent from the peak of 32,335 cases in May 1999.
3. During the same period, the share of unemployment cases among all CSSA cases dropped from 13.8 percent to 11 percent, and the ratio of unemployment CSSA caseloads to total number of unemployed persons in Hong Kong decreased from 15.5 percent to 14.8 percent.
4. On average, there were 637 unemployment CSSA cases closed per month during 1998/99. However, from June 1999 onward with the introduction of the SFS, on average, 1,067 unemployment cases were closed per month during 1999/2000, representing an increase of 67.5 percent over 1998/99.

These findings tend to support a first impression that the SFS is effective in curbing the growth of CSSA unemployment cases and is able to get unemployed welfare recipients back to work and helps them become self-reliant. However, closer examination of the captioned evaluation report also reveals the following concerns:

1. From June 1999 to May 2000, there were 7,232 unemployment cases being deregistered to participate in the AEA program. Among them, 1,794 people or 24.8 percent withdrew from AEA because they got gainful employment (getting a job with monthly earnings of at least HK$1,610). However, over half (61.7 percent) withdrew from AEA due to the following reasons: lost contact, withdrawal from applying CSSA, change in health condition/family commitment, died, and ineligible for CSSA due to assets exceeding prescribed limit.
2. The 1,794 successful CSSA job seekers only represent 9.5 percent of the cumulative enrollments of the 18,936 CSSA unemployed recipients participating in the AEA from June 1999 to May 2000.
3. As of May 2000, there were 2,282 CSSA unemployed cases participating in the community work (CW) program. However, only slightly over half of them (56.8 percent) were actually placed in CW projects. Besides, there were 2,350 deregistrations from CW during the period from June 1999 to May 2000, with more withdrawing cases (67.8 percent) from those people participating in environment-related CW projects than community services.
4. Analysis of variance and regression (ANCOVER) data show that the effect of the improved disregarded earnings (DE) provisions on success in job search is not obvious. Those who had been successful in regaining employment did not seem to have a more positive attitude toward DE as compared with those who remained unemployed.

The Hong Kong Council of Social Service also conducted an evaluation study on the SFS Scheme with seven focus-group interviews comprised of single parents and unemployed able-bodied persons, both males and females. These 32 participants did not find the SFS useful in helping them to get a job nor had the SFS been able to introduce a job to them. The jobs provided by the SFS staff were mostly outdated when they contacted employers for interviews. Worse still, it had increased each individual's psychological feeling of loss of face and sense of disgrace, especially in performing community work in environmental projects where each person is watched while cleaning country parks or beaches, thereby suffering the public stigma of a CSSA recipient. The following are verbatim accounts from some of these participants (Hong Kong Council of Social Service, 2001):

Middle-aged unemployed man: "Usually I waited for half a day but only got to see the worker in [the] Field Unit office for not more than five minutes. She only gave me two jobs on advertisements and asked me to report the result of my job interviews to them in a fortnight. There was no briefing on how to prepare for the job interview. I got the impression that the staff is only to check our job-interview record rather than teaching us how to get a job successfully."

Single-mother: "I became disappointed at the SFS. Staff there had not introduced me [to] any jobs. They only gave me advertised jobs and asked me to contact on my own for job-interview[s]. When I phoned, I often got the answer that the job vacancy was already filled."

Unemployed woman: "I had been warned seriously by staff in Field Unit offices that if I fail to comply with the required job-interviews, my CSSA payments would either be cut or terminated. And I had the experience of having half of my regular CSSA entitlement being cut."

Unemployed man: "The longer I participated in the SFS, the stronger is my feeling of uselessness and being an unwanted person in society. Look at us: more than forty years old, with only elementary schooling, and unwanted skills in unwanted trades. . . . The world is no longer ours. It belongs to young people. Every time I go to see staff at the SFS, I felt being checked upon for my performance. If I could have the choice, I'd rather rely on my own effort."

The Council's evaluation report (2001) commented that the SFS was not effective in the following ways:

1. Unable to help most CSSA unemployed recipients get a job successfully; job information provided was mostly outdated.
2. Unable to promote employability among CSSA recipients, especially for older-age-group people above forty. There was no job-interview training provided by field staff.
3. Neglected the difficulties faced by AEA participants, notably discharged prisoners due to an advancing age, personal background such as drug abuse, single parents, and chronic illness. The SFS scheme only pushes people to seek jobs.
4. Overoptimistic that jobs were readily available in Hong Kong though the economy was still in the downturn.

To conclude this discussion, the SFS was primarily a scheme aimed at getting people into work. The people who remained out of work faced life with a stigma attached, as well as social security benefits that were not adequate to meet their own or their families' needs, especially after assistance rates for able-bodied unemployment CSSA cases had been tightened in 1998. Structural factors and personal defects hindering their successful integration in the job market were not properly recognized by policy makers.

How Effective Are Reemployment Service Centers in Shanghai?

To cope with the large number of laid-off workers, Reemployment Service Centers were first set up in Shanghai in 1996. The service model is very different from the

Active Employment Assistance Programs in Hong Kong. By nature, these service centers were intermediate agencies set up by the Party Committee and Metropolitan Government of Shanghai and also the enterprises having laid-off and redundant workers. These centers are managed jointly by the Shanghai Labor Department and the enterprises. Every laid-off worker enrolled in these centers has to sign a trusteeship contract with an entrustment term of three years, during which time the centers gave them a basic living allowance; paid their pensions, medical, and other social insurance contribution fees; provided job training and retraining; and helped them in finding new jobs (*Beijing Review*, 18–24 May 1998; Mok and Huen, 1999; Chen, 2000).

Shanghai first experimented with these service centers to look after laid-off workers in the two traditional industries having the most laid-off workers, notably, textile and electronic appliances. Besides providing the above-mentioned support, a more proactive approach to promote reemployment was adopted:

1. Through the use of the Internet, training centers and job service agencies throughout Shanghai had been connected and kept informed of the training needs and market demands for new jobs (Mok and Huen, 1999). Bidding to offer these courses would be invited and eventually short-listed by the management of these centers and the Shanghai Labor Department, with training expenses to be borne by the latter. When the trainees finished their training courses, the training organizations and job service organizations would recommend them for employment; it was reported that in 1997 over 660,000 out of the 860,000 redundant and laid-off workers found jobs through the effort of these centers (Fudan University, 1998).

2. Enlarging the diversity of suitable jobs for reemployment. A good example is the case of the Shanghai Women's Federation which had successfully motivated the Shanghai Airline to loosen its age restriction for air hostesses to the age of thirty-six (Mok and Huen, 1999).

3. The Shanghai government had also created jobs in public services to absorb trainees from these reemployment service centers. The Labor Department in Putuo District had employed them to recycle waste and rubbish into papers.

4. Promoting the chance for self-employment. A representative case is the "Zhuang Mama Clean Vegetable Service" which employed female participants from these centers to buy, clean, and deliver vegetables to their customers who are mostly working parents.

Compared with the AEA program in Hong Kong, Shanghai's Reemployment Service centers tend to offer a one-stop range of cash assistance, training, employment assistance, and job-creation services. No wonder it was able to achieve high rates of reemployment and was adopted by the central government as a good model for other cities to follow. However, these centers also face the following problems:

1. Although the centers are able to provide a monthly basic living allowance to participants to a maximum of three years and also pay their medical and pension contributions, they offer nothing to help graduates who need to relocate to a new house. This happens to be a major problem for those workers in previous SOEs that provided them with living quarters. When they find a new job in another new enterprise, especially those working in private enterprises, and become private entrepreneurs, they would need to find a new house if the quarters did not belong to them. But the price for a new house is normally beyond their means. Thus, there is still a significant group of redundant workers being caught in this plight, who would appear very choosy for suitable jobs, and are thus not being successfully placed in jobs after the three-year entrustment period. Quite a number of workers would rather not join these reemployment service centers for fear of losing their housing, medical, and other corporate welfare as provided by their redundant SOEs.

2. The amount of cash allowance given by these reemployment service centers is very nominal, usually around 250 to 280 yuan per month. The rate for the second year is lower than the first year. Qiao (1999) observed that quite a number of trainees have to find some temporary jobs in order to feed their whole family.

3. Very few people eventually ended up starting their own business—less than 15 percent (*Shanghai Encyclopaedia*, 1999). Although some form of start-off incentive money would be provided by the Labor Department, most trainees would still prefer to be less risky.

4. When the three-year entrustment period expires and trainees are still not able to find a successful job, they can then be classified as unemployed and be referred for unemployment insurance for a maximum of two years. If they are unemployed for five years altogether, they can apply for the means-tested minimum-living-standard guarantee allowance, but the amount they would receive would be at basic subsistence level. The amount of cash benefits they would receive after a prolonged period of assistance would get smaller and more meager—a problem for those with young family dependents. Thus some redundant workers prefer not to join these reemployment service centers for fear of losing their employing status with their SOEs, as well as their eligibility to receive medical subsidy and other social security benefits. This is especially a problem when most hospitals now have to be self-financing and raise charges for hospital stays. Redundant workers would face difficulties when their operation and hospital expenses were not paid by their enterprises. Most workers still cast doubt

on their livelihood because reemployment service centers only promise to pay their medical charges and monthly basic living allowance to a maximum of two years. By then they would need to terminate their redundant status with their SOEs and become classified as unemployed workers. Thus some redundant workers would rather choose not to join these centers (Qiao, 1999). It seems that these centers are successful only for those participating laid-off workers and are ineffective for those redundant workers who choose not to join.

5. Reemployment service centers are funded equally by three sectors—central government, metropolitan government, and enterprises where such centers were created for their redundant workers (Chen, 2000). For near-bankruptcy and bankrupting SOEs, this would be another major drain on their limited capital and financial outlays. Some of them even have difficulties paying the medical bills for their redundant workers and would only promise a meager monthly basic living allowance to them.

Concluding Remarks: Modernizing Social Security in the Guise of a "Welfare-to-Work" Agenda

Toward the late 1990s, Hong Kong and Shanghai experienced major economic restructuring. Hong Kong, beset with the Asian financial crisis in 1997, faced an economic downturn and a rising unemployment rate to a record high level of 6.3 percent in 1999. At the same time, her CSSA unemployment cases continued to increase. Shanghai, under the central government's strong determination to modernize bankrupting and near-bankrupting SOEs into profitable enterprises, faced problems of massive layoffs and redundant workers despite a growth in her GDP by 10.2 percent in 1999. Correspondingly, there was a major growth in her unemployment insurance expenses and basic living allowance to laid-off and redundant workers. The two governments faced the same rising unemployment problem and mounting public expenditure on the dole and social security payments.

Policy makers in both cities were alert to the need to curb the growth of public welfare money spent on unemployment benefits. Thus social security schemes were repackaged with a Support for Self-Reliance motto for Hong Kong's CSSA scheme and an Active Labor Re-Employment Program among Shanghai's massive Reemployment Service centers. These programs were modernized with names of good intent—Active Employment Assistance Program (AEA), and Community Work in the case of Hong Kong, and Reemployment Community Service Centers in Shanghai. However, upon closer examination, Hong Kong's AEA program is only a job-advertisement introduction service and a job-interview checking system as reported by participants in these programs (Hong Kong Council of Social Service, 2001). There is little active employment assistance counseling to help unemployed CSSA persons cope with their personal barriers in job interviews, especially for middle- to older-age workers, single parents, and ex–drug addicts. Shanghai's Reemploy-

Table 6.7

Comparison of Hong Kong and Shanghai: Social Security

	Hong Kong	Shanghai
Main features:	Means-tested non-contributory in-cash benefit No explicit poverty line to determine benefit levels Benefit level ties to household composition and special needs (the special need component has been reduced since 1998) No cap on public expenditure in social security	Benefit level based on explicit poverty line with three different basic subsistence levels and multiple sources of funding Basic subsistence living being tax-based Insurance-based for the unemployment Basic wage borne by enterprises
Merits:	Benefit levels compared favorably with households in low-income employment Able to offer minimal standard of living Favors recipients with special needs	Explicit poverty line to determine benefit level Multiple funding source reduces the burden of the state Contributory Old Age Pension Scheme for retirement protection
Problems:	Income polarization and worsening underemployment create problem of work incentive and rapidly increasing public expenditure in this area	Benefit level too low Underfunded High unemployment, low insurance contribution, and solvent SOE affect functioning of unemployment insurance and basic wage system
Reform direction:	Improve work incentive by tightening the benefit level for working-age recipients and step up effort to help the unemployed recipients in job seeking	Increase workers' contribution rate for social insurance Step up effort for reemployment service (put to a halt recently)

ment Community Service Centers at times lose their attraction to those redundant workers who are afraid to lose their social security and medical and housing entitlement rights with their near-bankrupting SOEs. The monthly basic living allowance given to participating laid-off and redundant workers is low to meager. When the two-year entrustment period expires, and the workers lose out on their claim to unemployment insurance for another two years, these redundant and laid-off workers who could not find a job would become another group of new urban poor. During these turbulent years of economic restructuring, policy makers in both cities have done little to improve structural factors leading to massive unemployment.

Commenting on the experience of social security reforms under the new labor government in the United Kingdom, Grover and Stewart (2000) said that these reforms "are aimed at people in work or are aimed at getting people into work" (p. 248). Social assistance programs, when molded with this heavy work agendum, would tend to become "welfare-to-work" programs and would stigmatize welfare relief for those who do not want to work. Hong Kong seems to follow closely to this globalization path. The fate for Shanghai's unemployed and redundant workers does not appear better as they are increasingly faced with losing their "iron-rice bowl" to new forms of employment contracts on the eve of gradually surrendering their entitlement to enterprise welfare, housing, medical, and other social security benefits. Although new jobs probably offer higher pay, they are not secure jobs for life nor do they promise entitlement to living quarters. In transforming near-bankrupting SOEs to become profitable, there has not been enough attention paid to the gap that causes workers to lose their entitlement to housing, medical, and enterprise welfare. When new houses are expensive and when hospitals need to be self-financing and raise charges, life will be even more difficult, especially for the new urban poor of redundant and laid-off workers who are living on a meager monthly living allowance yet with a family and dependents to support. Unless structural factors leading to the rise in unemployment are dealt with, the prospects for the new urban poor and the nonworking CSSA recipients to escape from poverty are dim.

7

Elder Care

Alice Chong, Alex Kwan, and Gui Shixun

Population aging has become an important issue around the world. The size of the world's elderly population has been growing for centuries, but the rapidity of the present growth is unique. What is most troublesome is that both Shanghai and Hong Kong are aging at a speed that is greater than the rest of China and of the world, as shown in Table 7.1.

While most older people are active and contribute to society in various ways, a small proportion require assistance in daily living due to illnesses, frailty, or accidents (Ngan et al., 1996). The demand for long-term care will therefore accelerate with the continuation of population aging in both cities (Chong and Kwan, 2001). Long-term care is defined as "a broad range of services geared to helping frail older adults in their own home and community settings; [it] can include nursing homes, nutritional programs, adult day care, and visiting nurse services" (Hooyman and Kiyak, 2002, p. 360). This chapter aims to identify and compare the development of long-term care policies and direct service provision for older people who require assistance in their daily living in Hong Kong and Shanghai. It consists of four parts. Part one outlines the main characteristics of the aging population in the two cities. Part two introduces the social policies and welfare provision in relation to long-term care. Part three identifies the challenges that the two cities face in their attempts to provide care for the frail and demented older people. Part four makes recommendations for future developments in long-term care, drawing reference from the strengths and experiences of both cities. As there are different definitions of old age and no consensus has been reached, for the sake of clarity and comparison this chapter adopts the definition that is commonly used in the People's Republic of China, which refers to people of sixty years old or above.

Characteristics of the Aging Population

In the late 1970s, Hong Kong and Shanghai were among the first cities in China to become aging societies. According to the 2001 Census, Hong Kong had a total

Table 7.1

Aging Population (in Millions) from 2000–2050: A Comparison of the World, China, Shanghai, and Hong Kong

	2000				2030				2050			
	World	*China*	*Shanghai*	*HK*	*World*	*China*	*Shanghai*	*HK*	*World*	*China*	*Shanghai*	*HK*
Total population	6,080	1,261	13.25	7.12	8,139	1,483.12	13.45	8.7	9,140	1,479.47	11.88	7.55
No. of 60+	605	128.7	2.34	1.01	1,362	347.5	4.81	2.87	1,984	446.7	4.04	3.09
% of 60%	9.6%	10.2%	17.68%	14.2%	16.8%	23.4%	35.74%	33%	21.8%	30.4%	34.02%	39.8%
No. of 75+	150	27.42	0.6	0.3	386	92.83	1.39	0.87	707	187.02	1.43	1.55
% of 75+	2.3%	2.2%	4.53%	4.2%	4.7%	6.3%	10.36%	10.1%	7.8%	12.7%	12.07%	20%

Sources: Data for the world, China, and Hong Kong: U.S. Bureau of Census, International Data Base (Oct. 5, 2000), "Midyear Population, by Age and Sex," Retrieved on March 31, 2001, from www.census.gov/cgi.bin/ipc/idbagg. Data for Shanghai: Working Group on "Review of Shanghai Social Security for the Aged: System and Operation," *Review of Shanghai Social Security for the Aged: System and Operation* (Shanghai: Shanghai Scientific Publication Publisher, 1998), p. 64.

population of 6.72 million people; 1 million (14.9 percent) were aged sixty years or over, and 0.29 million (4.3 percent) were aged seventy-five or over (Census and Statistics Department Homepage, 2002). Shanghai had a total population of 13.27 million people, which was double that of Hong Kong. Among that population, 2.46 million (18.58 percent) were aged sixty and above, and 0.32 million (2.5 percent) were aged eighty and above (Shanghai Research Center on Aging, 2002).

Great Speed in Population Aging

Both Hong Kong and Shanghai have seen an aging of their population in the past few decades, and will witness an even more tremendous trend in the future. In Hong Kong, people aged sixty and above comprised only 6 percent of the population in 1966, but this increased significantly to 10 percent in 1981, making Hong Kong an aging society. The figure then rose to 13 percent in 1991 and 14.9 percent in 2001. It is projected to increase to 25.9 percent in 2029 (Census and Statistics Department Homepage, 2002) and to 39.8 percent in 2050 (U.S. Bureau of Census, International Database, 2000).

The aging population comprised 6.1 percent of Shanghai's total in 1964, and this increased to 10.1 percent in 1979, when Shanghai became an aging society. From 18.6 percent in 2001 (Shanghai Research Center on Aging, 2002), the figure is projected to rise to 32.3 percent in 2030 (Shanghai Municipal Committee on Aging and Shanghai Research Center on Aging, 2000), and 34 percent in 2050 (U.S. Bureau of Census, International Data Base, 2000). It is estimated that Shanghai's aging population will peak between 2025 and 2030 (Shanghai Research Center on Aging, 2002). Hence, by 2050 four of every ten citizens in Hong Kong and three-and-a-half of every ten citizens in Shanghai will be sixty years old or more.

In the same vein, the absolute numbers of older people in both cities are overwhelming. Hong Kong had about 1 million older people in 2001, a number that is projected to rapidly increase to 2.3 million by 2029 (Census and Statistics Department Homepage, 2002) and to 3.1 million in 2050 (U.S. Bureau of Census, International Database, 2000). There were 2.4 million older people in Shanghai in 2001 (Shanghai Research Center on Aging, 2002); this figure is projected to rise to 4.8 million in 2030 and to decline gently to 4 million in 2050 (U.S. Bureau of Census, International Database, 2000). Given its larger population size, Shanghai will have to support even more older people than Hong Kong in the coming decades, although Hong Kong's percentage of older population will catch up and exceed that of Shanghai by the 2040s.

This great increase in aging populations is mainly due to improved longevity (Chong, 1996). The Chinese life span has increased at a dramatic rate in the last fifty years. Life expectancy at birth in Shanghai was only 58.3 years for men and 60.3 years for women in 1953. Due to the cessation of warfare, more equitable land and food distribution, reductions in malnutrition, improved water supplies, and a general increase in and emphasis on hygiene and sanitation (Arnsberger, Fox, Zhang, and Gui, 2000), there has been a tremendous improvement in the quality of life. Life expectancy approached 71 years for men and nearly 75 years for women in 1978. The figures became 75 years for men and 79 years for women in 1998, which were very similar to the average life spans in developed countries of 71.2 for men and 79.2 for women (Shanghai Municipal Committee on Aging and Shanghai Research Center on Aging, 2000).

The life expectancy at birth in Hong Kong was 75.2 years for men and 80.7 years for women in 1991 and 77.2 years for men and 82.4 years for women in 1999. The figure is projected to increase to 82.3 years for men and 87.8 years for women in 2031 (Census and Statistics Department, Homepage, 2002). Indeed, the United Nations has ranked Hong Kong third in the world in longevity, with an average life span at birth of 79.5 years, closely following Japan (81 years) and Sweden (79.7 years) (*Ming Pao*, 25 July 2002).

The great speed of population aging is, in fact, a positive indicator of the modernization and industrialization of both cities, as it reflects significant improvements in sanitation, health care systems, and quality of life. For the first time in centuries, the Chinese saying that "it is rare to be 70 years old" is no longer true for people living in Hong Kong and Shanghai. While we should rejoice in the good news, we should not neglect the needs of a small proportion of the older generation who encounter difficulties in their daily living (Kwan and Chan, 1999a; 1999b). Among them, many are the old-old—those aged seventy-five or above.

Large Proportion of the Old-Old

At present, a great majority of the older adults in both cities belong to the young-old, that is, they are aged below seventy-five. However, the old-old group will grow at a faster rate than any other age groups in the future. In Shanghai, those who were aged 75 or above were estimated to number 0.6 million in 2000, representing 4.5 percent of the total population. This number will increase to 1.39 million (10.4 percent of the total population) in 2030 and to 1.4 million (12.1 percent of the total population) by 2050. In Hong Kong, those who were aged 75 or above occupied 4.2 percent of the total population in 2000; this number is expected to increase sharply to 10.1 percent in 2030, and will continue to grow to 20 percent by 2050 (U.S. Bureau of Census, International Database, 2000). These figures suggest that there is likely to be a greater demand for long-term care in the coming decades. Moreover, the demand will be even more serious in Hong Kong than in Shanghai due to its larger proportion of the old-old.

Social Welfare for the Older People

Historical reasons such as differences in ideology, financing, and governance between Shanghai and Hong Kong mean that there were differences in the ways that older people were cared for in the two cities in the past. However, convergence has occurred in the last few years with the increase in contacts between the two cities.

While the use of the term "long-term care" is gaining headway in Hong Kong, it is very seldom used in Shanghai, where "social welfare" is current. The following section first highlights the social welfare system of each city, how policy is formulated, and how resources are distributed to provide long-term care to older people who are frail or demented. It then delineates the various welfare services that serve the elderly. In both cities, community care and family care are the major focuses of care for frail older people. Both cities also recognize the need to expand institutional care.

The Social Welfare System in Shanghai

In mainland China, social welfare is part of a broadly defined social security package that consists of social insurance, social assistance, welfare services, compassionate rehousing, mutual aid, and individual savings-accumulation systems. According to the 1992 White Paper titled "A Report on the Development of China's Social Welfare Services" (Ministry of Civil Affairs, 1993), social welfare services are special protective measures for the material and cultural life of the people. They include various social welfare institutions, community services in urban areas, preferential treatment and resettlement of demobilized servicemen and the families of martyrs, social welfare lottery sales, welfare enterprises and the employment of the disabled, social insurance in rural areas, and social relief for the victims of natural disasters. Social welfare services represent the efforts of the state, collective units,[1] and work units to safeguard the basic living standard of the people through the provision of material assistance, welfare facilities, and welfare services. Social welfare serves not only older people but also children, youth, women, the disabled, and ex-servicemen and their families, as well as the rural population. In terms of services, social welfare provides institutional facilities, community services, social insurance, and disaster relief and resettlement, as well as employment services.

In 1982, the National Committee on Aging advocated five focal points for services for older citizens: achievement of a sense of security, appropriate medical care, a feeling of worthiness, continuous learning, and a sense of happiness (*Encyclopaedia of Chinese Social Work*, 1994). A very high-level policy-making mechanism, the National Commission on Aging was established in October 1999 under the State Council, which is the highest policy-making body in China. This reflects the Chinese government's high level of concern about aging issues.

Compared with other cities in Mainland China, Shanghai started early in recognizing the challenge of an aging population. It established the Shanghai Panel on Aging in 1991. This was later promoted to the status of the Shanghai Municipal Committee on Aging, which is a high-level policy-making and coordinating body. It is chaired by one of the deputy secretaries of the Shanghai Municipal Party

Committee, and two Shanghai deputy mayors act as the vice chairmen. Other members include representatives from each governmental department and unit of the Shanghai Municipal Government.

Among various government departments, the Ministry of Civil Affairs has been given the main responsibility to develop social policies for the care of older people and be responsible for service provision. A Bureau of Civil Affairs exists in each city; its Department of Social Welfare (previously known as the Department of Civil Affairs and Welfare Work) has the responsibility of taking the lead in providing and monitoring various welfare services, including services for older people. As shown in its 2001 Annual Report, the major focus of the Department of Social Welfare of the Shanghai Bureau of Civil Affairs is older people. Of the thirteen major tasks that were noted in its 2001 Annual Report, eight of them involved elder care, especially institutional care (Department of Social Welfare, Shanghai Bureau of Civil Affairs, 2002).

Older people also enjoy legal protection of their human rights in China. "The People's Republic of China Rights of the Elderly Ordinance," which was formally put in place in 1996, specifies the guiding principles of protecting older people's basic rights and the different duties and responsibilities of the state, various government structures, social enterprises, the family, and the elderly. One special feature of the ordinance is the requirement for older people's children and other relevant people to provide their elders with financial, physical, and spiritual assistance and support. In response to this, Shanghai enacted the "Shanghai Municipal Rights of the Elderly Ordinance" in 1998, which requires the different levels of municipal government to plan and implement various welfare facilities for the elderly. These ordinances help to regulate services for the older generation and to provide better safeguards for them.

The Social Welfare System in Hong Kong

In Hong Kong, "social welfare" as an umbrella term encompasses a wide range of services, including social security, family and children services, medical social work, probation service, rehabilitation services for people with disabilities, services for the elderly, youth services, and experimental projects with disadvantaged groups such as new immigrants. However, there is no special ordinance to safeguard the human rights of the elderly (Kam and Kwan, 1996; Kwan, 2000). Under the principle of "One Country, Two Systems," Chinese legal ordinances are not applied in Hong Kong.

In Hong Kong, the chief executive, together with the Executive Council, and in some instances the Legislative Council, formulates overall welfare policy, including long-term care policy. When important policy initiatives are involved, the government first prepares a green paper or consultation document for public consultation; the final policy directives are then issued in the form of a white paper or program plan. Various governmental departments serve as executive arms and set policy objectives into operation. Among various government departments, the Social Welfare Department shoulders the greatest responsibility in the provision of long-term care for the elderly. It holds the primary responsibility for setting welfare goals into operation, allocating new services, financing and monitoring services, and service coordination. It only provides a few direct services—mainly social security to people with various financial needs, probation services, family services, and medical social work. The majority of welfare services are delivered by approximately 180 nongovernment organizations under the financial subvention of the Social Welfare Department.

"Caring for the elderly" has been one of the Strategic Policy Objectives outlined by the chief executive, Tung Chee-hwa, in his annual policy address since 1997 when Hong Kong was returned to China. The government has pledged to provide its older citizens with a sense of security and belonging, as well as a feeling of happiness and worthiness. To show his concern, the chief executive appointed an advisory body, the Elderly Commission, in July 1997, with representatives from the Health, Welfare and Food Bureau and from government departments such as the Social Welfare Department, the Health Department, the Hospital Authority and the Housing Department. Academics, professionals, and social leaders are also invited to participate on an individual basis. The commission's objectives are to formulate a comprehensive policy for elder care, to coordinate the planning and development of services, to recommend priorities, and to monitor the implementation of policies and programs for the elderly.

Types of Welfare Services in Shanghai

Before 1978, the various levels of government provided few formal services for older people in Shanghai. There was no policy and few resources were allocated for care of the older generation. Only a few welfare homes and "homes of respect"[2] were established, mainly for the "three-nos" (no family support, no work ability, and no means of livelihood) elders, that is, older people who had no family members, no dependable relatives, and no income.

A change of government policy occurred in 1978 when China adopted a socialist market economy to achieve the policy objectives of the Four Modernizations. The Civil Affairs Administrative units at different levels of government were then restored. While serving the three-nos elderly was still the priority, a change of emphasis occurred in the financing and provision of welfare services: from a closed-door to an open-door policy in terms of financing, from an emphasis on charity and basic needs to a welfare and holistic orientation in terms of service provision, and from a care orientation to a care-and-rehabilitation orientation (Chong, 1996). From this point onward, services for the elderly gradually grew with the support and leadership of Civil Affairs Administrative Units at different levels of government, including municipal government, district government, street and residents' associations in the urban areas, and counties', townships', and villagers' associations in the rural areas.

Compared with other cities, Shanghai has put much effort into providing care for older people. The services, including both community care and institutional care, are provided or financed by the various levels of government, collective units, enterprises, and individuals.

Community Services in Shanghai

The development of community services in Shanghai was prompted by a meeting on national community service that was held in Wuhan in 1987. In the meeting, the Chinese State Civil Affairs Ministry recommended the development of community services in major cities to serve people with special needs, including older people. In response to this, the first street-level community service center for the elderly—Dapu Community Service Center—was established in Luwan District in Shanghai. Other centers have subsequently been established in different streets. Although the size of these centers varies greatly, they usually provide services such as canteens, cafés, activity centers, drop-in centers, bathing areas, laundries, rehabilitation services, barbershops, repair and maintenance services, and telephone hot lines. A few centers even provide institutional care. Another initiative is that of neighborhood nursing care groups, which are set up by street offices and which rely on volunteers from among retirees, local officials, cadres, neighbors, and students to provide assistance to older people without family members. The volunteers may help in daily purchases, meal preparation, hospital visits, the escorting of older people to medical consultations, or care for the sick. Very often, an alarm bell is installed at the older person's bed to permit them to contact their neighbors in case of emergency. At the same time, local groceries, stores, hospitals, coal shops, and so on deliver their services or products to the homes of the needy elderly.

In 1991, the Shanghai Community Services Committee—a high-level body for policy formulation, coordination, and monitoring—was established. It was renamed the Shanghai Community Services Coordinating Committee in 1996 and is responsible for planning and coordinating all community services in Shanghai, including those for older people. In 1992, the Shanghai Municipal Government identified the expansion of services for the elderly as one of the government's twelve strategic tasks for the year. In 1994, the government proposed planning standards in community services, including a few for the care of older people. These include the following:

1. A community service center should be established in each street or township, 95 percent of residents' associations should set up a sub-center, and no less than 12 percent of the total population in an area should be involved in voluntary work that is related to the elderly.
2. Government departments should make provisions for the planning of street-level community service centers in their regular planning exercises and should provide resources to establish and operate such centers.

3. The subsidy to community service centers from various governmental structures should be no less than 20 percent of the total cost of running such a center.
4. Public enterprises, community organizations, collective units, overseas Chinese, and individuals are encouraged to run community service centers.
5. Municipalities and districts are encouraged to establish trust funds for developing community services. Tax reductions are also provided for donations to community services (Gui, 2003).

An important policy paper, "Shanghai Municipal Development in Elder Care—The 9th Five Year Plan and 2010 Vision Objectives," was formulated in 1997. One of its objectives was to establish an elderly rehabilitation center and a social center in each residents' association in urban areas and in each township in rural areas. The paper also identified the need to provide meal delivery, bathing, and hair-cutting services to the needy elderly. A note of caution is that though these policy directives are formulated with good intentions, the extent to which they are implemented depends very much on the priorities and resources of individual levels of government structure. Nevertheless, in general the quantity of service units and facilities has greatly increased in the last decade.

In January 2000, there were ten district-level community service centers. Out of 120 streets, 115 street-level community service centers were established, and out of the 2,820 residents' associations, 2,441 subcenters were established. There were also 1,500 activities centers and 2,787 "families of respect for the elderly."[3] Alarm clocks have been installed for 2,809 elderly people living alone. Other services include meal delivery services, escort services for medical consultation, manual assistance, and support groups for elderly people with special needs. A citywide hot line service has just been established (Gui, 2003). These statistics show that, overall, the increase in in-home and community care facilities has been quite significant and the coverage is high. By the end of 2001, all nineteen municipal districts had established steering committees on community care and had operated community service centers and hot lines. Streets and counties (55.87 percent or 138 of all) had set up their own service units. Altogether, 4,790 elders were receiving community care from different levels of government. Among them, 22 percent attended day-care centers, and 78 percent received in-home care (Department of Social Welfare, Shanghai Bureau of Civil Affairs, 2002).

Institutional Care for Older People in Shanghai

In China, the main types of institutional services for senior citizens are welfare homes, homes of respect, and hostels.[4] Welfare homes at the municipal level are totally financed and operated by the state under the administration of the Civil Affairs Bureau. All other institutions may have mixed forms of financing. Initially, these aged-care homes were intended to serve the three-nos elderly. However, with the adoption of the socialist market economy by the 1978 Third Plenary Session of the Eleventh Central Committee of the Chinese Communist Party and the subsequent open door policy (Chong, 1996), self-financed older people are also accepted, such as retired people with pensions and elderly people whose family members cannot take care of them. Besides elderly people, the municipal welfare homes also serve disabled people, mental patients, and orphans.

In 1992, the municipal government formulated a policy to establish at least one infirmary in each district or county to tackle the great shortage of infirmary beds for elderly people who were suffering from higher levels of impairment. In 1994, the Shanghai Bureau of Civil Affairs advised that at least one aged care home (either a welfare home, a hostel for the elderly, or a home of respect) should be built in each district level. The district should sponsor no less than 20 percent of the construction costs and the costs of equipment and furnishing. The homes' incomes would be exempted from profit and investment tax. Moreover, charges for water, gas, and other utilities were to be set at discretional rates.

The "Shanghai Municipal Development in Elder Care—9th Five Year Plan and 2010 Vision Objectives" document of 1997 also recommended increases in the provision of institutional care. In particular, it proposed increasing the total number of institutional beds from 12,621 in 1995 to 22,000 in 2000 and to no less than 40,000 in 2010. Besides the increase in total numbers of beds and institutions, the plan also specifically recommended improvements in the quality of service of the municipal level welfare homes and the construction of welfare homes in counties and districts that can afford them. The document outlined the target for expansion at each level of government. For example, it recommended that each district or county construct two to three welfare homes and add no less than 4,000 beds by the year 2000. At a lower level of government, each street in urban areas or township in rural areas was asked to establish at least one or two homes of respect, with the aim to add no less than 6,000 beds by the year 2000. Residents' associations and village committees were recommended to set up at least one home for the elderly. The document also recommended the involvement of individuals and social organizations, which were expected to provide 5,000 beds by 2000.

Recognizing the need to differentiate between policy formulation and the execution of policy, the Shanghai Bureau of Civil Affairs has decided that its Department of Social Welfare will focus on policy formulation and monitoring. Meanwhile, a new Shanghai Municipal Welfare Center was established in 1998 to oversee the construction of new institutions and to be responsible for their day-to-day operation and management.

To ensure the proper management of institutional care, the "Shanghai Municipal Management Guidelines for Institutional Care" have been in force since 1 October 1998. The guidelines recommend, first, that resources should be raised through various channels, such as donations in kind and in cash, bank loans, sales of welfare lotteries, and the provision of voluntary service. Second, institutional care should be part of economic and social development at each level of government. Third, institutional care should be divided into two classifications: nonprofit and for-profit. Charges for the former are fixed by the government, while providers of the latter can choose between following the fees that are fixed by the government or deciding on their own fees. This gives greater flexibility to for-profit institutions to generate more resources. Fourth, nonprofit aged-care homes should enjoy tax exemptions and discretionary charges for various utilities. Fifth, the operators of all homes should arrive at service agreements with residents' family members to specify the kinds of service to be provided and the fees involved. To promote service quality, homes for the aged should keep health records and provide regular health checks and rehabilitation activities for residents. Sixth, community hospitals in the area should provide outreach services to residents in aged-care homes. Seventh, experts and professionals should be formed into teams to assess the service quality, facil-ities, resources, and agency image of each aged-care home. Being broad and gen-eral, these guidelines leave much room for maneuver by facility operators. However, they do point out the right direction for aged-care homes to develop and operate.

As of the end of 2001, there were 453 institutions providing 31,163 beds for senior citizens. Of these, 22,778 beds (73 percent) were provided by different levels of government. They were divided among 25 municipal welfare homes, county or district welfare homes, and 322 street-level homes of respect. Another 8,385 beds (27 percent) were distributed between 106 homes of respect and elderly hostels that were run by community groups. All of these beds only served 1.3 percent of the older people aged sixty and above in Shanghai (Department of Social Welfare, Shanghai Bureau of Civil Affairs, 2002). Apart from homes that were under the civil affairs system, in January 2000, 1,784 beds were in twenty-seven infirmaries, which are operated by the Health Care Bureau (Gui, 2003).

Types of Welfare Services in Hong Kong

Even though the provision of welfare service for the frail elderly commenced earlier in Hong Kong than in Shanghai, it had a late start when compared with welfare services for other target groups, such as families and youths (Kwan, 1997, 2002). Development in elder care quickened after the launching of the White Paper on Social Welfare—"Social Welfare into the 1980s"—in 1979. The White Paper as-serted the importance of community and family care; for the first time, the gov-ernment agreed to shoulder responsibility for expanding community support services for senior citizens. Moreover, the government also acknowledged the need to increase the provision of institutional care for older adults who could not take care of themselves or whose family members could not take care of them.

During the 1990s, recognizing the importance of old age support in the context of rapid aging, the government (both before and after the return of Hong Kong to Chinese rule) placed great emphasis on the care of older people. The 1991 White

Paper on Social Welfare, "Social Welfare into the 1990s and Beyond," took the innovative step of using the aging population instead of the general population as the basis of the planning ratio for new services units. For example, it was recommended that social centers for the elderly should be established in each neighborhood with 3,000 older people aged sixty or above, and that multiservice centers should be established in districts with 25,000 people aged sixty or above. It also explicitly acknowledged the need to expand various services for older people. Finally, it recommended linking the fees that were charged for elderly services to their cost in order to expand resources.

The Working Group on Care of the Elderly was appointed by the Hong Kong governor in 1993 to review existing services and to make recommendations for the development of various social services for older people, including welfare services, housing, medical care, education, and social security. Its report in 1994 recommended, among other things, standardization for planning purposes of the definition of "old age" as sixty-five years or above. All the planning ratios for different services were adjusted accordingly. In terms of service development, the report suggested the expansion of home help services, the setting up of a central waiting list for residential care for the elderly, and the launching of outreach geriatric and psychogeriatric teams. In terms of financing, the report encouraged self-financing and privatization.

As laid down by the 1991 White Paper "Social Welfare into the 1990s and Beyond," the broad objective in providing services to the elderly is to promote the well-being of people aged sixty or above in all aspects of their lives, to enable them to remain members of the community for as long as possible, and, to the extent necessary, to provide residential care that is suited to their varying needs. Therefore, services for the elderly can be classified into community support services and institutional care services, just like those of Shanghai. However, recognizing the need to provide a continuum of long-term care to older people with different levels of impairment, a number of new initiatives have been introduced since 1999 to achieve a better integration of social, nursing, and medical services. Consequently, the separation between community care and institutional care is now less distinct.

The Social Welfare Department (SWD) provides long-term care services mainly through subvention to non-profit-making, nongovernment organizations (NGOs).

Community Services in Hong Kong

Community services are provided to enable older people to remain in the community for as long as possible and to provide assistance and support to family caregivers. These services include social centers (which mainly provide social and recreational activities); multiservice centers (which provide a wide range of services including counseling, home help, social and recreational activities, community education, networking for older people, bathing, laundry, and canteen services); daycare centers (which provide personal, nursing, and paramedical care for the senile

Table 7.2

Publicly Funded Community Services for the Older People in Hong Kong (as of 30 September 2002)

Type of service	Number
Neighborhood elderly center/social center	216
District elderly community center multi-service center	37
Day care center	41
Day respite service (pilot project)	12
Day care center for demented elderly (pilot project)	4
Home help team	139
Home care	25
Meal service	25
Enhanced home and community care service	18
Holiday center for the elderly	1
Outdoor and recreational bus service	4
Career support center	2

Sources: Hong Kong Council of Social Service, *Information on Services for the Elderly in Hong Kong—2002* (Hong Kong: Hong Kong Council of Social Services, 2002) (in Chinese); Social Welfare Department, *Social Welfare Department-Service Sectors: Services for Elderly*, retrieved June 20, 2003, from www.info.gov.hk/swd/html_eng/ser_sec/ser_elder/index.html.

aged); home help services (providing meal delivery, personal care, and escort services to older people living in the community); home care (which provides in-home services); meal services (which only provide meal deliveries); and one holiday camp for the elderly and four coaches that transport older people with mobility difficulties on outings or visits. The statistics on all of these services are detailed in Table 7.2.

In the last few years, quite a number of initiatives have been launched to help elderly people with special needs (e.g., frailty or dementia) stay in the community and to prevent premature institutionalization. First, several initiatives have been launched to support informal caregivers of the older people. The Hong Kong government has financed two support centers for carers since 1999. These centers provide a range of direct services, training and educational programs, and promote mutual help among informal caregivers of the elders. A few other multiservice centers have also launched carers' projects, some with financial assistance from the Jockey Club and others from the Community Chest. A pilot three-year project on day respite services came into operation in October 1999, and twelve day-care centers participated, each serving three places. One main objective of the project has been to allow informal caregivers to have a break. Second, efforts are made to cater the special care needs of demented elders. A pilot project for carers of the demented elderly has been launched, and four-day care centers provide specialized training and programs to twenty demented elders each, as well as relief to and support for their family carers. Third, in-home care has been significantly strengthened. In 2000, eighteen Enhanced Home and Community Care Service teams were established to provide various in-home services for elderly people with different levels of impairment. This service aims at filling the service gap that is left by existing home help services, which provide a variety of personal care services (meal delivery, personal care, feeding, assisting in bathing, housecleaning, escorting to medical consultation, and so on) but are not capable of dealing with the specialized medical and nursing needs of the frail and demented. Advice and assistance from various human service professionals, including social workers, physiologists, occupational therapists, nurses, and even doctors are used to provide holistic and integrated care to individual elders. Moreover, as of April 2003, existing home help teams are financed to upgrade their service into integrated home care service in order to serve the frail or demented elders as well. Fourth, a revamping of the community service was launched in April 2003, and multiservice centers and social centers for the elderly are encouraged to restructure their services to form district elderly community centers (DECC) and neighborhood elderly centers (NEC), respectively. Their major service is to serve the elders with special needs, such as those who are frail or demented and those who are suffering from different social and psychological problems. They are also expected to provide support to the informal caregivers, and to promote the health of the elders.

Institutional Care for Older People in Hong Kong

Different types of institutional facilities are financed by the government to meet the needs of older people who are experiencing different degrees of frailty or dementia. In ascending order in terms of the degree of care provided, they are hostels (mainly for those who are capable of self-care, but they are phasing out); aged-care homes (providing meal services); care-and-attention homes (providing personal care and limited nursing care); nursing homes (serving those who require regular nursing and medical care); and infirmaries (serving those who require active rehabilitation, medical, and nursing care). The number of places that are provided by the various types of institutions is shown in Table 7.3. The total capacity of all these institutions has not yet reached 30,000 beds, and they serve less than 3 percent of the older people in Hong Kong. This capacity is severely insufficient, and there has always been a long waiting list for institutional care. For example, 26,029 older people were on the central waiting list for residential care services as of 31 May 2003; three-quarters of these required regular assistance in personal care or nursing care, as they had applied either for care-and-attention homes (65 percent) or nursing homes (21 percent). The average waiting time for publicly funded care-and-attention homes amounted to thirty months (Social Welfare Department, 2003).

Due to the serious shortage of publicly funded aged-care homes, the provision

Table 7.3

Number of Publicly Funded Institutional Beds (as of 31 March 2003)

Type of service	Number
Hostel	166
Home for the aged	7,431
Care-and-attention Home	11,356
Bought place scheme and enhanced bought place scheme from private aged care home	5,841
Nursing home	1,484
*Infirmary	2,555
Self-financing home	2,903
Emergency placement	145

Sources: *D. Tsang, Financial Secretary, *The 2001–2002 Budget of the Hong Kong Special Administrative Region Government,* retrieved on March 31, 2001, from www. info. gov. hk/budget01-02/bgtshf.html; Social Welfare Department, *Social Welfare Department–Services Sectors: Services for Elderly,* retrieved June 20, 2003, from www.info.gov.hk/swd/html_eng/ser_sec/ser_elder/index.html.

of private aged-care homes has become a booming industry in Hong Kong since the mid-1990s. According to the director of social welfare (5 March 2001), approximately 40,000 beds (68 percent of all aged-care home beds in Hong Kong) are provided by the private sector. The Social Welfare Department has resorted to purchasing beds from better-quality private aged-care homes under the Bought Place Scheme and the Enhanced Bought Place Scheme[5] to shorten the waiting period for applicants.

However, due to their profit-making nature, the quality of private aged-care homes varies greatly. The Residential Care Homes (Elderly Persons) Ordinance was introduced on 1 April 1995, to regulate the operation of both publicly funded and for-profit aged-care homes through a licensing system. The ordinance specifies requirements for building and fire safety, space standards, and staff ratios for aged-care homes providing different levels of care. Moreover, to promote the service quality of private aged-care homes, a one-off Financial Assistance Scheme was introduced in June 1995 to help private and self-financed aged-care homes renovate their buildings to reach the standards that were required by the ordinance.

Besides these types of aged-care homes, some new initiatives have also been introduced in the past few years. First, twenty-nine infirmary units have been set up in nineteen care-and-attention homes, each serving twenty infirm residents. An infirmary allowance has also been provided to other institutions to help them employ additional staff to take care of residents who are infirm and on the waiting list for infirmary placement. The introduction of these measures recognizes the fact that some residents require a level of care that is higher than that provided by existing institutions. Second, a three-year pilot project has been introduced to serve the elderly suffering from dementia. Since 1999, six Dementia Units have been established in five care-and-attention homes, each serving twenty-four to twenty-eight demented residents (Social Welfare Department, 2001). Third, continuum of care is being promoted in the aged homes. All new institutions are required to adopt such an approach so that residents with increased physical or cognitive impairment can remain in the same institutions.

Besides the Social Welfare Department, the Hospital Authority, which is in charge of all the publicly funded hospitals in Hong Kong, also provides some form of long-term care. It operates community-based geriatric assessment teams and psychogeriatric assessment teams, which use an outreach approach to provide assessments of older people who are suffering from poor physical and mental health. Infirmaries have been established to provide acute rehabilitation and medical care for the elderly, while day hospitals provide rehabilitation to individual elderly patients who are newly discharged from hospitals or whose health problems do not require admission into hospitals.

Challenges

Although Shanghai and Hong Kong have begun to lay out policy directives and develop various long-term care services for their frail senior citizens, both cities still have to face a number of challenges before their older people can age in dignity. These challenges include the increasing demand for care, the weakening of the ability of families to take care of the elderly, and the severe shortages in service provision.

Increasing Demand for Care

Estimating the number of elders who require assistance in their daily activities is difficult. Arnsberger et al. (2000) have listed twenty predictors for Chinese older people, such as being seventy-five years old or older, living alone, having two or more chronic conditions, requiring assistance in two or more activities of daily living, having no children or having no children living nearby, and so on. Generally, demographic changes play an important role in increasing the demand for care.

As previously mentioned, both cities are projected to experience great increases in their aging populations and in the proportion of the old-old. While these demographic changes reflect improvement in the societies and should be celebrated, they also mean that there is an increasing number of older people who require some assistance in their daily living. In addition, the percentage of older people who are living alone or living with other elders is changing, which may increase the demand for long-term care.

Living alone has repeatedly been found to be one of the best predictors of institutionalization in the United States (Arnsberger et al., 2000). Yet, living alone or with only a spouse is overrepresented among the older generation in both Hong Kong and Shanghai. In 2000, 10.2 percent of those aged sixty or older in Hong Kong lived alone, and 20.1 percent lived with their spouses. The figures were much higher than those for middle-age people who were aged from forty-five to fifty-nine, which were 3.7 percent and 9.8 percent, respectively (Census and Statistics Department, 2001). In Shanghai, 8.6 percent of households were occupied by single elders, and 34.8 percent by single elders and elderly spouses in 1998 (Kwan, Gui, and Chong, 2003). With the current urban renewal plan, under which old neighborhoods are being redeveloped, many of the elderly are being moved into high-rise apartments far from their children (Arnsberger et al., 2000). At the same time, the rise of a private property market in Shanghai will lead to more and more young couples moving to new apartments and leaving their parents with "empty nests." The number of empty-nest elderly families rose from 28 percent in 1986 to 34.8 percent in 1998, and is projected to increase to 65 percent in twenty to thirty years' time (Kwan, Gui, and Chong, 2003). At present, empty nests may not have negative effects on elderly parents who are living alone. A survey in 1998 found that among the present cohort of elders, 54 percent had 3–5 children and 24 percent had 6–9 children. Hence, even if the elderly live alone, it is still likely that some of their children will live in the proximity and provide help. However, the effect of empty nests will be serious after 2010, when the first single-child parents start to enter old age.

One of the authors conducted research with associates on the elderly population in Shanghai during 1998, in which 3,524 older people were interviewed. The group found that 7.5 percent of participants had problems in self-care. Among these participants, 5.1 percent self-assessed themselves as partially capable of self-care, while 2.4 percent were totally incapable of self-care. Among those aged eighty and over, the need for care was even greater, as 25.9 percent were only partially capable of self-care, and 9.9 percent were completely incapable of self-care (Shanghai Municipal Committee on Aging and Shanghai Research Center on Aging, 2002). A Task Force on Aging Issues in Shanghai also made a similar estimation that about 6 percent of older people were incapable of self-care (Ketiju, 1998). These data have two implications. First, the percentage of older people who require assistance in their daily living is significant, and their caring needs may be aggravated if they live alone. Second, the old-old have a greater tendency to require long-term care. Taking care of elderly dependents, then, will present a great challenge to the caring capacity of the family, the community, and the state.

Weakening of the Caring Ability of Families

Traditionally, frail Chinese elders were taken care of in their own homes by their family members. It was commonly assumed that the spirit of filial piety would provide strong incentive for adult children to take care of their senile parents. Yet, can the family shoulder the increasing caring task?

A look at the statistics shows a shrinking of family size in both cities in the last fifty years. In 1950, the average family in Shanghai contained 4.6 people; by 1960, the number was 4.5; in 1970, it was 4.8; and in 1980, it lowered to 3.8 people. After the implementation of the one-child policy in the 1970s,[6] the average family size dropped to 3.1 in 1990 and 2.9 in 1998 (Gui, in press). The typical family structure has changed from that of a large extended family with many children and grandchildren to one with few children and grandchildren. Indeed, since the launch of the one-child policy, a new family structure, the "421 family," has emerged. This means that a young adult couple has to take care of four parents and parents-in-law in their sixties, as well as one child. Indeed, with increases in longevity, the couple may have to take care of their old-old grandparents in their eighties as well, thus increasing the number of dependent old-old family members to a maximum of eight. The family structure will then become "8421." The sheer number of dependent elderly family members may overwhelm the caring ability of any filial couple.

Apart from the reduction in family size, the caring ability of the family has been weakened by three other factors. The first factor is an increase in the workforce participation rate of women. In the past, family care in China meant care by daughters, daughters-in-law, and wives. Nowadays, nearly all adult women participate in the workforce. For example, the 1982 Census found that 90 percent of women in the twenty to forty age range participated in the workforce (China National Committee on Aging Problems, 1988). Hence, there has been a sharp reduction in the supply of unpaid female caregivers at home. Second, many informal carers are spouses or children who are usually quite old themselves, and it is difficult for them to shoulder the caring responsibility. Third, the caring needs of elders who are moderately or severely impaired or demented require more than patience and personal care. Very often, carers need to have a proper understanding of impairments such as dementia and stroke before they can accept the elders' behavior and provide appropriate care. Otherwise, family carers can be at a loss, anxious, and under stress.

A similar situation exists in Hong Kong. There is no one-child policy in place, but the caring ability of families has also been weakened. This is due to an increase in the workforce participation rate of women (from 47 percent in 1993 to 49 percent in 1998) and an increase in the percentage of nuclear families to total households (from 59.2 percent in 1986 to 63.6 percent in 1996) (Chu, 2000).

These data clearly show a "care gap" (Walker and Walker, 1985) between the demand for family care and the supply of unpaid family caregivers. With shrinking family sizes, the increase in the aging population will exert great pressure on medical care, social welfare, housing, and income security.

Shortages in Service Provision

There are acute shortages in the provision of community and institutional care in both cities, especially in the service of older people with higher levels of impairment.

In terms of statistics, Shanghai may not seem to suffer a shortage in institutional care. There has been a great increase in the number of beds for the elderly from 20,203 in 1998, to 28,454 in 2000, and to 31,163 by the end of 2001 (Department of Social Welfare, Shanghai Bureau of Civil Affairs, 2002). The ratio of beds for those aged sixty or above increased from 0.86 percent in 1998 (Shanghai Municipal Committee on Aging and Shanghai Research Center on Aging, 2000), to 0.93 percent in January 2000 (Gui, 2003), and to 1.3 percent in 2001 (Department of Social Welfare, Shanghai Bureau of Civil Affairs, 2002). Moreover, the utilization rate for homes under the Civil Affairs Bureau is only around 70 percent (Gui, 2003). However, a shortage in institutional care has been repeatedly discussed by the Shanghai Municipal Committee on Aging and Shanghai Research Center on Aging (2000). Moreover, the 1.3 percent of beds for those aged sixty or above is much lower than that of other industrialized societies. For example, the U.S. Bureau of the Census revealed that 4.3 percent of the population who were aged sixty-five or above were receiving some kind of institutional care in 1998 (Hooyman and Kiyak, 2002). This paradox is mainly due to the decrease of the three-nos elderly from the introduction of pension systems by enterprises and the state. Additionally, some older people cannot afford the cost of institutional care, while others might find life in aged-care homes too monotonous and that the service quality and facilities do not meet their expectations (Gu, Lin, Lu, Chu, and Zhu, 2000). For example, not all aged-care homes have emergency bells, and very few offer couples' rooms. Hence, the demand for such homes is underreported. Moreover, the provision of institutional care for those who are infirm or bedridden is extremely insufficient. The extent of this shortage is most likely to be underestimated, as some families do not send their frail members to aged-care homes because of financial constraints.

In addition to insufficiency in supply, institutional care in Shanghai also suffers other problems, including great variations in the quality of care, insufficient resource inputs, poor management, and the misuse of facilities and resources (Shanghai Municipal Committee on Aging and Shanghai Research Center on Aging, 2000).

Hong Kong presents a much worse situation. Even though the total number of institutional beds that are financed by the government can serve nearly 3 percent of the older population, which is 1.7 percent more than that in Shanghai, this number is still much lower than that of other industrialized societies. Even though the utilization of aged-care homes is high, ranging from 94 percent to 97 percent (Tsang, 2001), the shortage is still very serious. Due to the tremendous demand for institutional care, there has been a waiting list of around 30,000 applicants for different types of institutional beds for several years. Therefore, quite a number of the applicants die while awaiting admission. For example, it is estimated that 2,500 applicants for care-and-attention homes die every year (*Ming Pao*, 6 June 2000). Despite the enforcement of the Residential Care Homes (Elderly Persons) Ordinance, the service quality of aged-care homes that are run by for-profit organizations still varies greatly, thus discouraging their use. Apart from shortages in institutional care, a similar problem exists in community care, though to a lesser extent. There are long waiting lists for services such as day care and home help.

Recommendations

The peak of aging in both Hong Kong and Shanghai has yet to come. Hence, the coming decade will provide enough lead time to formulate relevant policies and to establish the infrastructure that will be needed to face the real challenges that are projected for after 2010. As previously mentioned, the reductions in family sizes, the rising percentage of elders who are living alone, and the high workforce participation rate of women have weakened the ability of the family to perform the traditional role of taking care of its frail members. Filial piety is a noble spirit among the Chinese, but it needs to be backed up by appropriate in-home support and other long-term care services to assist adult children and other family members to care for their frail and aged relatives for long periods of time. However, even now there are severe shortages of formal care in the two cities. Unless more proactive measures are taken, the challenges that are identified in the last session will be aggravated.

In view of the tremendous speed of population aging and the challenges that were mentioned previously (as summarized in Table 7.4), ten recommendations are now proposed for policy makers and stakeholders who are concerned about the welfare of our older generation. These suggestions range from policy suggestions to direct service options, cover the role of the government, modes of financing, and the mobilization of different resources in the care of the older generation.

Proactive Policy Making

The Shanghai Municipal Committee on Aging provides an excellent forum for policy formulation, as all of the relevant government departments and bodies are involved, and their representatives are of sufficient rank to ensure the smooth implementation of service initiatives. The Committee should make good use of these strengths and formulate long-term and proactive social policies. Given the fact that the aging population will not reach its peak in Shanghai until 2025–2030, there is still time to lay out the infrastructure to face the related issues. These include constructing a continuum of long-term care facilities, training staff, conducting community education to instill proper attitudes about old age and old people, conducting research to identify the older generations' needs and expectations, and soliciting resources for these services.

In comparison, the Elderly Commission of Hong Kong is not as powerful. It is only an advisory body, and does not possess the authority and leadership of the Shanghai Municipal Committee. Even though the commission has authorized many studies and has made many recommendations, its contribution to policy formulation

Table 7.4

Summary Table on Comparison of Population Aging and Long-Term Care between Shanghai and Hong Kong

	Shanghai	Hong Kong
Characteristics of aging population	Became aging society in 1979 Great speed in population aging: 18% in 2000 to 34% in 2050 75+ overrepresented: 5% of total population in 2000 to 12% in 2050	Became aging society in 1981 Great speed in population aging: 14% in 2000 to 40% in 2050 75+ overrepresented: 4% of total population in 2000 to 20% in 2050
Policy and planning in elder care	Elder care a high priority Special national ordinance to protect elders' human rights Shanghai Municipal Committee on Aging: a high-level policy-making and coordinating body Several policy papers in last decade but no systematic planning standards for services Department of Social Welfare of Bureau of Civil Affairs: mainly responsible for service monitoring and provision Institutional care serving 1.3 % of the elders Great variety in government financing in different districts and streets	Caring for the elderly has been a strategic policy objective of chief executive since 1997 Elderly Committee: advisory in nature only Last policy paper written in 1994 Policy directives and planning standards for most services The Social Welfare Department (SWD): responsible for planning, financing, service monitoring, and coordination Institutional care serving 3% of the elders Pluralism in financing is encouraged
Service delivery in elder care	Majority of elder care provided by different levels of government structure Community care is the focus Great increase in service provision in last decade Staff usually without relevant training Community care relies heavily on volunteers and informal cares	Majority of elder care by NGOs with heavy subsidy from the SWD Community care is the focus Wide range of community and in-home care options Many new initiatives and experimental projects Moving toward a continuum of care Many staff are professionals
Challenges	Increasing demand for care from improved longevity, empty nests, and living alone Weakening of family caring ability (one-child policy resulting in 421 family structure and great rate of female participation in the workforce) Shortage in service provision	Increasing demand for care from improved longevity and living alone Weakening of family caring ability Great shortage in service: long waiting list for institutional care

is rather limited. Very often, the Health, Welfare and Food Bureau, and the Social Welfare Department only adopt parts of its recommendations. Consequently, reforms and policy initiatives appear to be fragmented and piecemeal. To make the commission effective, its terms of reference and authority should be strengthened along the lines of its Shanghai counterpart. Moreover, some of the research studies that are contracted out by the Elderly Commission or by the government are undertaken by overseas consultants, who may lack comprehensive understanding of the local situation and culture. Hence, we strongly recommend the inclusion of local researchers or resource personnel in consultancy teams. Finally, the last policy paper by the Working Group on Care of the Elderly was issued in 1994; another strategic policy paper on care of the elderly is overdue. In the policy formulation process, the voice of the elderly and their informal caregivers should be heard and taken seriously.

The Role of Government

In the past, the Department of Social Welfare of the Civil Affairs Bureau in Shanghai was involved in both policy formulation and the daily operation of municipal welfare homes, thus creating role conflicts and a lack of accountability. In the 1990s, the department separated policy formulation from daily operation. The department is now responsible for policy formulation and service monitoring, and the Municipal Social Welfare Center is responsible for the construction of institutions and the day-to-day management of services. We recommend that the department focus its efforts on the macro level, including formulating vision and policy objectives for care of the elderly, providing monitoring, and initiating legislative change. We also recommend that the department encourage the establishment of non-profit-making welfare organizations and that it finance them to provide the actual services. In such a manner, not only will community participation be enhanced, but it would also be likely to lead to greater cost-effectiveness than the current practice whereby the majority of service delivery is provided by different levels of government.

Unlike their counterparts in Shanghai, the Hong Kong Social Welfare Department (SWD) and the Health, Welfare and Food Bureau have very clear ideas of their roles in policy formulation, coordination, and service monitoring. As has been the case in other industrialized societies, such as the United States, the United Kingdom, New Zealand, and Australia, a series of welfare reforms was introduced by the Hong Kong government at the end of the 1990s. These reforms included the establishment of a Service Performance Monitoring System in 1999, the allocation of lump sum grants in 2001 to finance NGOs, and the introduction of competitive bidding for new service allocation in 2001. Among these new initiatives, the Service Performance Monitoring System has contributed a great deal to monitoring the quality of welfare services. It includes the following features: First, a "funding and service agreement" is signed between the Social Welfare Department as the funder and the NGOs as the service providers, spelling out their respective roles and responsibilities, and the output (including both quantity and quality of service) that is to be provided by the NGOs. Second, a set of sixteen Service Quality Standards has been established, and all service units, including both governmental and nongovernmental units, are required to adapt and apply these standards in their own service units and to develop operational guidelines for service provision. These sixteen standards cover the rights of the users, the service provision process, and the management of service units. Third, each service unit is required to carry out annual self-assessment on its ability to meet the sixteen standards, as well as the service requirements that are specified in the Funding and Service Agreement. Each unit has to submit a self-assessment report to the Social Welfare Department each financial year. Moreover, the unit is required to prepare an improvement plan should it fail in any area of assessment. At the same time, a special unit has been set up by the SWD to conduct an external assessment of each service unit every three years. Service units that fail the assessment will be given suggestions for improvement, and persistent failure may lead to the termination of government subsidies.

Admittedly, the introduction of these welfare reforms in Hong Kong is a step in the right direction. However, they were launched within a short period of time and without sufficient consultation, especially on the issue of competitive bidding for the allocation of new services. The reforms have thus created many uncertainties and confusion in the welfare sector. Very often, NGOs that need to rely on government subsidies have no choice but to comply with the new rules. However, an even more significant effect of these reforms is on the role of NGOs. In the past, the NGOs that provided the majority of welfare services were treated as partners by the government. Consultations and negotiations were often held before the introduction of major changes in the welfare field. Now, made more confident by its recognition of the power it gains from holding the welfare purse, the government has become very directive: It formulates policies without consulting service providers or the elderly and then announces and implements them. Its position is "take it or leave it." There are always other organizations, including profit-making organizations, that are only too eager to run the services. While the government may be delighted by its newly gained power and apparent improvements in cost-effectiveness, this change in the relationship between the government and NGOs will inevitably destroy the biggest contribution of the NGOs, namely, their role as advocators and innovators of services, as well as partners in policy formulation. We therefore recommend that more dialogue be held between the government and service providers to promote mutual understanding and to reduce the gap between written policy and actual service delivery.

Our second recommendation is to learn from other countries' experience. As many industrialized societies embarked on the welfare reforms earlier than Hong Kong, we recommend reviewing the achievements and limitations of the reforms in these countries. It is always better to look before you leap.

Our third recommendation is to have an overall blueprint for care of the older generation in the coming decades before actually launching new policies. At present, many new initiatives become outdated within a very short period. For example, the home care scheme that was launched in 1999–2000 in Hong Kong was replaced by the Enhanced Home and Community Care Scheme in 2001–2002. The interface between these two services remains blurred. NGOs spent considerable resources in the design of home care services, only to find that they would no longer exist the next year and that they had to design yet another new service.

Our fourth recommendation is that experimental projects should be tried before large-scale implementation. This strategy will achieve better utilization of resources and have fewer negative effects on service development.

Pluralism in Financing

In view of the great demand for services, the different levels of government in Shanghai should shoulder greater responsibility for establishing and financing service units for older people. This support might take any of the following forms. First, it might take the form of financial assistance for the construction and/or the daily operation of service units. Financial assistance is especially important at the construction stage, as the capital investment that is required may be too substantial for community groups or enterprises. Second, it might take the form of better discretionary policy (such as discretionary utility charges or tax exemptions) toward non-profit-making enterprises and associations that establish or provide services for the elderly. Third, it might take the form of securing support from other governmental departments or bureaus for the allocation of land, the approval of architectural plans, or the securing of bank loans at discretionary rate.

The quasi-governmental welfare homes created a different set of problems. The Civil Affairs Bureau has plans to require quasi-governmental welfare homes to become self-financed, and only provides financial assistance to beds for the three-nos elderly. Financing long-term care for the elderly may be a great burden on any government. While it is proper to aim for pluralism in financing, the pace of the changeover cannot be too fast; otherwise, the cost will be transferred directly to users, which will lead to huge service fees or residents being forced to withdraw from services due to financial constraints. Rapid changeovers may also lead to substantial layoffs of staff, affecting overall staff morale and ultimately the service quality of institutions. Therefore, we recommend that resources be provided by the different governmental structures at a ratio that although not as high as the present financing level, is in proportion to the increase of the old-old. At the same time, welfare homes should be encouraged and assisted to achieve better cost-effectiveness (through measures such as cutting staff costs and increasing capacity), to try self-financing or even profit-making tactics to generate additional resources, and to target a proper mix of profit-making, self-financing, and state-financed beds.

In Hong Kong, the government has shouldered the greatest share in funding welfare services and has found this to be too great a burden. Therefore, it has tried different ways to contain costs and increase productivity, such as the introduction of competitive bidding for new service units, the launch of lump sum grants to provide flexibility to service providers in the use of financial resources, and the Enhanced Productivity Project. Besides public money, the Community Chest and the Lotteries Fund have provided some financial support in the construction of facilities and in subsidizing experimental projects. Different trust funds and charitable organizations have also provided funding support to help people in need, including older people. In contrast to the situation in Shanghai, service providers in Hong Kong should not rely too heavily on government subsidies but should explore new and creative ways to generate revenue, as well as to maximize the efficient use of resources.

Continuation of Care

Ideally, older people should be able to "age in place"; in other words, they should be able to age in an environment that is familiar to them, without having to adapt to substantial changes in places of residence or service providers as their physical or mental competence declines. In both cities, the majority of institutional care facilities provide service to older people with specific levels of self-care abilities. When their health deteriorates to a more serious level, they have to be transferred to other institutions or centers that provide a higher level of care. This relocation or change of service providers may cause great adjustment problems among the elderly and their caregivers. Therefore, in 2000 the Hong Kong government instituted two experimental projects, one in institutional care and the other in community care. The former provides additional resources to three participating aged-care homes to meet the caring needs of more senile residents. The latter is the "Enhanced Home and Community Care" project, which provides a wide range of long-term care service options to frail older people in the community. In 2003, existing home help teams are encouraged and financed to upgrade their service to serve elders with moderate or severe impairments. These kinds of efforts are certainly steps in the proper direction and should therefore receive greater emphasis.

The continuum of care should range from in-home service to institutional care, providing different levels of personal, nursing, and medical care. It might even include hospices. A crucial issue is the interface between the welfare and medical sectors. In Hong Kong, this interface implies integration, closer cooperation, and mutual respect between social workers under the auspices of the Social Welfare Department and the doctors, nurses, and health care professionals under the Hospital Authority and the Department of Health. Although the Social Welfare Department, the Department of Health, and the Hospital Authority are all under the same Health, Welfare and Food Bureau, there is much room for improvement in coordination and cooperation among them. The situation is even worse in Shanghai, where the Civil Affairs Bureau and the Bureau of Health are independent and separate entities.

We recommend that an experimental project on long-term care with close cooperation between the medical and welfare sectors be carried out in a particular district in both cities to find the best ways to achieve integration between the two systems. This may involve the exchange of users' files; team approaches in need assessment and formulating intervention plans; the involvement of staff/professionals who are working in different community services, aged-care homes, medical, and health facilities; support for carers; and the use of information technology to facilitate proper record exchanges and communication between different service providers. Moreover, a case management approach should be adopted so that the same professional case manager will continuously assess the needs of an individual older person and match them to the appropriate services according to their degree of frailty. With the increase in types of services, it will be impossible for individual older persons or family members to identify different services or to differentiate between the functions and nature of services.

Another direction in terms of aging in place is to strengthen community and in-home care over institutional care, because the majority of older people prefer aging in their own homes to aging in institutions. Indeed, the 1998 Shanghai Older Population Integrated Survey found that 96 percent of participants expressed such a preference (Shanghai Municipal Committee on Aging and Shanghai Research Center on Aging, 2000). This has two implications. First, more innovative approaches should be encouraged to identify a better mix of different social, nursing, and medical care to help elders with different levels of impairment and dementia in the community and in their own homes. Suggestions and feedback from different professionals, family caregivers, frontline formal carers, and the elderly themselves should be continuously solicited. Second, resources that were originally planned for institutional and hospital care should be transferred to community care. Very often, community care is adopted by many countries under the rhetoric of more individualized care, better care quality, genuine concern by carers, and so on. Yet, the main motive is to cut costs. Although there may be some savings involved in the use of community care, mainly from the reduction of premises and the administrative cost of operating round-the-clock aged-care homes, the proportion that is saved is not great. Community care is by no means cheap if comprehensive and intensive care is to be provided to moderately or severely impaired elders in their own homes, which usually calls for the combined efforts of various professionals and may even require a twenty-four-hour care plan for some severely impaired elders.

Reducing Service Shortages

In view of the acute shortage of services in both cities, we recommend the expansion of service provision in the coming decade to face the peak of the aging population after 2025. Shanghai has so far heavily relied on volunteers and community groups, such as community nursing care groups, to provide community care. This kind of community support network should be further promoted as it provides a timely and caring service for those who are living alone or for three-nos older people. However, the informal support network will be overwhelmed if it is required to shoulder the full responsibility for the care of a rapidly growing aging population. Formal community support services must therefore be planned and developed as soon as possible.

Three suggestions should be kept in mind when expanding service provision in Shanghai. First, instead of arbitrarily setting quotas for different services, the utilization of resources would be maximized by adopting a ratio that uses the older population as the basis of planning, for example, the number of welfare home beds per 1,000 older people. In this way, districts with a higher proportion of older residents will be provided with proportionally more relevant resources and services. Second, while there is a general need to expand all services for the aged, it is necessary to focus on services that meet the long-term care needs of the frail and demented elderly, as they are the groups who have the highest demand for care and who severely tax the caring abilities of informal caregivers. In other words, resources that are currently reserved for aged-care homes that provide a lower level of care, such as hostels and homes of respect, should be redirected to long-term care facilities that provide a higher level of care. Third, in the planning and launching of new services, provisions should be made to allow room for future conversion to provide a continuum of care and a higher level of care to the increasing number of frail and demented older people. In Hong Kong, many aged-care homes have to undergo large-scale reconstruction before they are architecturally suitable for the provision of care-and-attention home service. These modifications are not only costly, but also are highly inconvenient to existing residents, who may need to be temporarily transferred to other institutions.

In the case of Hong Kong, two recommendations can be made. The first is to speed up the provision of institutional facilities. The government has rightly been trying different tactics to increase the provision of institutional care for infirm and demented older people. These include the expansion of the Enhanced Bought Place Scheme from private aged-care homes; asking aged-care homes to convert into care-and-attention homes; adding beds in care-and-attention homes; encouraging the establishment of self-financing and profit-making institutions; and the provision of dementia or infirm allowances to aged-care homes. However, the increase is still far too slow for the long waiting list. It is true that the government has considered other long-term care alternatives such as the Enhanced Home and Community Care Service, partly to overcome service shortages, partly to save resources, and partly to enable older people to stay in their own homes for as long as possible. As mentioned, we support a greater emphasis on community care over institutional care. However, due to the long waiting list for institutional care and the fact that some frail elders and family members of demented elders prefer institutional to community care, the government still needs to speed up the provision of aged-care home beds to meet the pressing needs of some of those on the waiting list.

Second, even though the government wants to control welfare expenditure, additional resources should be bestowed upon services for older people in proportion to the increase in the aging population. The government has launched several new initiatives to save resources or to add value to resources, such as competitive bidding for new service projects, the Enhanced Productivity Project which requires every service unit to save 5 percent of its total expenditure in three years' time, and encouraging more services to be provided on a self-financed or profit-making basis. However, due to the rapid expansion of the aging population, and especially of the proportion of the old-old, additional resources are definitely needed to meet the ever-expanding service needs of the aged.

Promoting Service Quality

Although there have been many new initiatives in Shanghai, welfare services for older people are still at an initial stage of development. Some services are financed by profit that is generated by the commercial operations of the Street Offices (Leung, 1996). The quality of these services varies significantly, depending on the availability of resources and the leadership of individual Street Offices. Many services are loosely structured and lack standardization, with service options differing substantially from district to district and from street to street. In comparison, most welfare services in Hong Kong have planning ratios and are guided by policy directives. The publicly funded services enjoy standardization in staffing ratios, in operational subsidies, and in the allocation of premises, furniture, and equipment. Moreover, the Service Performance Monitoring System helps to promote service quality among subvented services. The only weakness is that the system lacks performance indicators to demonstrate the quality of service that is provided, but the welfare sector has been taking action to formulate such performance indicators for different long-term care services. A more pressing quality issue in Hong Kong lies with the private sector since its service quality is still highly uncertain and unstable.

While a few institutions in both cities have voluntarily gained ISO accreditation to demonstrate their good service quality, this only represents the effort of a small proportion of aged-care homes. In what follows, recommendations are made to promote the service quality of long-term care.

First, we recommend that the governments of both cities devise comprehensive service monitoring systems to ensure service quality. Such systems will require the formulation of performance indicators that represent service quality as experienced by service users, as well as by their family carers. At present, the Hong Kong government is very keen on introducing clinical performance indicators, such as records of falls and incontinence, in the newly allocated service units. These indicators may be appropriate for institutional care but not for community care, as they neglect the psychosocial aspects of care and do not have enough space for direct feedback from users and their family members. However, the Hong Kong government has appointed, in 2002, the Hong Kong Association of Gerontology to conduct a two-year consultancy study on an accreditation system for all aged-care homes in Hong Kong. This should also be considered by policy makers in Shanghai.

Second, staff quality and performance should be strengthened. Younger staff of better education and caliber should be recruited and trained in Shanghai to provide professional care, as well as to take up managerial and supervisory roles in the care of the elderly. At present, many management personnel are retirees. Due to their advancing age, they may only work for a few years and then truly retire, leading to high staff turnover. The majority of the senior management is made up of cadres who have no prior work experience in long-term care services. Consequently, many of them are at a loss to find ways to provide proper care to the elderly and to manage the centers or aged-care homes.

Indeed, in many industrialized societies, working with older people is not a popular job for younger people. This situation is especially serious in Shanghai because, unlike Hong Kong, Shanghai does not consider welfare services professional in nature. The social status of social welfare workers is low, salary and fringe benefits are meager, and the work is tiring and sometimes tedious. It is therefore very difficult to attract well-qualified and competent staff. We suggest that welfare services for the elderly be identified as professional services to be undertaken by trained social workers and paramedical professionals. Corresponding increases in salary and fringe benefits will be required to make such careers worthwhile. As to the existing staff, they should be given time and support to upgrade their qualifications through the attendance of professional courses.

Moreover, systematic in-service staff development programs should be organized for all levels of staff in both cities. The need for training is especially strong in Shanghai. At present, very few staff members have received any formal or preservice specialized training. Many of them have prior work experience in areas that are vastly different from welfare services. Some might have been farmers, and some might have been cadres of the Communist Party. Even though the Ministry of Civil Affairs has arranged for municipal welfare homes to organize in-service training for operators and personal care workers, these staff development programs are mostly short and not comprehensive enough. The focus of such programs should be the dissemination of knowledge, the development of proper attitudes, and the acquisition of skills (including personal care skills, skills in planning and running different programs and activities, counseling skills, and management skills). At the same time, principles and ethics for working with older people, such as empowerment, respect, and self-determination, should also be instilled. Finally, we suggest that different training plans should be tailored to the various training needs of staff members with different job responsibilities.

Third, older people and family members of the frail elderly in both cities should be empowered to speak up and to express their expectations and assessments of service quality. In Hong Kong, older people usually wait for two to four years before they gain admittance to aged-care homes. This, together with their relatively low self-esteem, makes them hesitant to voice criticism. Therefore, they must be actively encouraged to give feedback, including negative feedback. At the same time, service operators should respond to comments as soon as possible to show their respect for their users' views.

Shanghai has another concern in promoting service quality, that is, how to make it work. The city has both the bureaucratic structure and the assessment mechanisms to promote service quality. The separation of policy formulation from service execution in Shanghai has provided the required structure for monitoring service quality. In theory, now that the Department of Social Welfare of the Civil Affairs Bureau is mainly responsible for policy formulation, financing, and monitoring, it is in a better position than in the past to monitor the service quality of the operating service units. In practice, however, few aged-care homes or centers are truly NGOs. Most of them are affiliated with lower levels of government and the staff members are practically civil servants who enjoy job security and comprehensive fringe benefits. Hence, it is difficult to motivate them. These service units should be privatized to become semipublic enterprises in order to allow greater flexibility in the deployment of resources. Government subsidies should be continued, maybe at a decreasing rate as with the welfare subvention system in Hong Kong, and financial self-sufficiency should be encouraged.

In terms of assessment mechanisms, Shanghai has introduced methods for monitoring the service quality of institutional care since the mid-1990s. In 1994, the Civil Affairs Bureau issued the "Temporary Guidelines on Management of Welfare Homes," which specified the charging of fees according to the frailty of residents, the establishment of residents' associations, and the promotion of six responsibilities and five basic rights of the residents (including the rights of religion and traditions of ethnic minorities). In 1996, the Civil Affairs Bureau revised the assessment methods for aged-care homes in Shanghai (Gui, 2003). However, the assessment mainly focuses on medical care and the management of the resources and administrative procedures of the institutions and neglects the psychosocial health of the residents. Moreover, feedback from residents is not systematically solicited. The "Shanghai Municipal Management Guideline of Institutional Care" that was introduced in 1998 added, among other things, the establishment of service agreements between aged-care homes and their residents or residents' family members. All these measures are steps in the right direction. However, the establishment of a monitoring system alone is not sufficient; it will not work unless all stakeholders, including operators, staff, residents, family members, and officials take it seriously. This requires a cultural change that will take time; otherwise, existing practices will prevail. There is a common saying in China: "Where there is a policy from above, there are always counteracting strategies from below."

Support for Informal Caregivers

Traditionally, the majority of older people in both cities are taken care of by their family members, mainly women. As previously mentioned, there is a serious weakening in the caring ability of the family caregivers. Hence, training and support for informal caregivers should be a high priority in both cities.

First, training should be provided to family carers in understanding the common illnesses of older people and ways to help frail or demented elders toward rehabilitation. Second, concrete support services should be provided. These include respite services, day-care centers, day hospitals, and palliative care. Counseling and stress management training should be made available to informal carers who are facing great anxiety or are under stress. Of all services, respite care, both day and residential, is especially important in sustaining informal carers, as they can easily burn out if not given a break from time to time. Moreover, such carers usually feel tied down by their duties and might decline any offer of service if no respite is provided to take care of their frail or demented dependents when they go out. Third, mutual help should be fostered among family carers. The sharing of experience between informal caregivers can provide mutual support and much-needed advice.

In Hong Kong, the Social Welfare Department has financed two resource centers for carers, while a few multiservice centers have launched carer projects using different sources of financing. Yet, the services are still far from adequate, and caregivers may have to travel long distances for help. However, the Hong Kong government has just launched a new initiative of revamping the community support services. A major feature is to ask the multiservice center to expand its service to become the district elderly community center (DECC) in March 2003, and to serve the frail and the demented, as well as to support the informal caregivers. Therefore, more resources and higher priority should be given to supporting informal carers. This need is even greater in Shanghai, where support for carers has not yet received the emphasis that it deserves.

Mobilizing Social Support Networks

The mobilization of volunteers and neighbors to help older people has been more successful in Shanghai than in Hong Kong. The concept of the "voluntary contract care provider"[7] may provide a good example to Hong Kong of how this can be done. Even though special staff members have been allocated to each district-based multiservice center in Hong Kong to promote volunteer service, the volunteers' services are mainly short term (for periods such as three months) and infrequent (such as once a month). Many provide psychological support, such as phone contact or home visits, but not basic personal care. A much greater effort in mobilizing, training, and rewarding volunteers is needed in Hong Kong. In 1998, the first author visited Japan, which boasts the world's greatest longevity. She was impressed by its "welfare bank" system, whereby the service hours of each volunteer are recorded and returned to them later when they are in need. These hours can also be transferred to the volunteer's loved ones. This becomes a great motivator for volunteer service. Efforts along similar lines should be considered in Hong Kong and Shanghai.

An Ounce of Prevention Is Worth a Pound of Cure

The majority of older people are capable of self-care; only those who suffer from various illnesses or disabilities require assistance in their daily living. Health promotion should therefore be a top priority in any aging society. In both Hong Kong and Shanghai, the emphasis in the care of older people has been more on treatment and less on health promotion. It is recommended that equal emphasis should be placed on health promotion, illness prevention, early detection, early treatment, and rehabilitation.

Legal Protection

Legal protection of older people's basic human rights can foster a positive environment. This is what Mainland China did in 1996 when it put into force "The People's Republic of China Rights of the Elderly Ordinance." While making it mandatory for adult children to take care of their elderly parents will require more deliberation in Hong Kong, other parts of the ordinance can provide very useful references in safeguarding the basic human rights of the older generation.

Conclusion

Both Shanghai and Hong Kong have recognized the potential challenges of population aging, and have already started to do something about them. Comparatively speaking, policy on care for elders in Hong Kong is in place, resources input from the government is more substantial, and services are better institutionalized and more abundant than those in Shanghai. Nevertheless, even though it had a later start than Hong Kong, Shanghai is very determined to tackle aging issues, and great progress in policy formulation and the expansion of service provision has been seen in the last few years. It provides the older generation with better legal protection of their basic human rights and can mobilize more social support networks in serving elders who are in need. Therefore, the two cities have much to learn from each other. Exchanges, site visits, and discussion forums should be organized for policy makers, elders, family carers, and professionals (including social workers, nurses, doctors, and paramedical professionals). Moreover, in view of the expected rapid increase in the absolute number and percentage of the old-old in the coming decades, strategic plans should be formulated to meet the long-term care needs of the frail and demented elderly in both cities.

Notes

1. Collective units include street and neighborhood communities in urban areas, as well as townships and villages in rural areas. They are the lowest level of the government structure, and are responsible for developing public education, social services, cultural activities, and public services at the grassroot level.
2. Welfare homes are built and administered by the Civil Affairs Bureau at the municipal/district/county levels to serve residents of different age groups and different self-care abilities, including the elderly, orphans, cadres, and patients with disabilities. They are few in number but generally possess better manpower and resources than the other institutions. Homes of respect are run by various collective units and work units at the level of the street or residents' associations. They mainly serve those who require low levels of care. The service capacity, resources, and service quality vary greatly among different homes.
3. Such families are formed when a desolate older person invites a few other desolate elderly people to move into his/her own housing unit, so that they can keep each other company. Another option is that they can gather together under one roof in the daytime, and return to their own housing units at night. These older people are mainly capable of independent living. The number of elderly people who are involved can range from two to several.
4. The hostels provide small, self-contained housing units (i.e., with kitchens and toilets) for single older persons or elderly couples who are capable of independent living.
5. The Enhanced Bought Place Scheme is for older people who require higher levels of care than are provided by the Bought Place Scheme.
6. To control the increase of the nation's population, the Chinese government introduced a series of family planning mechanisms in the 1970s, including the single-child policy whereby each married couple is only allowed to give birth to one child. However, due to strong resistance from rural areas and minority groups, rural and ethnic families are allowed to have a second child.
7. This is usually promoted at the township or street level. It mainly serves the three-nos elderly, elderly family members of martyrs and soldiers, and the retired elderly who are in need. Through signed contracts, individuals, neighbors, residents, village committees, or multiple service units agree to provide specified services at specified times to a particular older person, usually in the neighborhood. The services can include the delivery of food and utilities, washing of clothes, medical care, house cleaning, escorting to medical consultations, and even funeral arrangements.

—————— 8 ——————

Migration and Competitiveness

Ray Yep, King-lun Ngok, and Zhu Baoshu

> International migration can be said to be a good thing on the basis of revealed preference: employers hire foreign workers, and workers migrate to jobs abroad.
> —Charles Kindleberger, quoted in Hollifield, 1992, p. 45

Based on his observation of the trajectory of economic development in postwar Europe, Kindleberger's assessment of the economic impact of cross-border labor mobility is highly positive. Immigration emerged as a central feature of state-led strategies for economic growth in Europe during this period (Hollifield, 1992). Anxious to recover from destruction and destitution, the Federal Republic of Germany had recruited some 12 million persons during the 1950s. In France, average annual immigration also jumped from 66,400 in the period 1946–1955 to 248,800 in the period 1956–1967. The shortage of domestic workers and the cheapest labor costs account for the extensive use of immigrant workers. As a new country two centuries ago, America also used endogenous resources for rapid development. As early as 1791, Alexander Hamilton had warned the American Congress of the necessity of encouraging imported labor if the "scarcity of hands" and "the dearness of labor" were to be avoided. A century later, Andrew Carnegie went further by framing the economics of migrant workers in explicit monetary terms: "[T]he value to the country of the annual foreign influx is very great indeed . . . these adults are surely worth $1500 each—for in former days an efficient slave sold for that sum." In Carnegie's phase, immigration was "a golden stream which flows into the country each year" (Calavita, 1994, p. 56). In the global age, following the emergence of global markets, the growing laxity of border control, and on-line distribution of employment information, international labor mobility has further accelerated. By 1999, an estimated 130–145 million people lived outside their countries, up from 104 million in 1985 and 84 million in 1975. Corollary to this brain drain is the rocketing of undocumented migration. Trafficking has become a booming business, moving perhaps 4 million people a year, generating U.S.$7 billion (United Nations,

1999, pp. 32–33). These waves of legal and illegal migration have provoked serious reflections on policies of citizenship and naturalization across the world (Aleinikoff and Klusmeyer, 2000).

What is evident from this cursory review of migration trends over the last two hundred years is the inevitability of cross-border mobility of labor and the potential economic contributions of these migrant workers. Uneven economic development and opportunity across countries appears to be a perpetual magnet attracting the steady movement of personnel. This justifies some forms of state regulation and restrictions, necessary for the preservation of sovereignty and effective governance in the recipient state. However, the quantitative and qualitative contributions of migrant labor, past and present, should never be overlooked. Importing skilled and cheap labor can be an effective and economic way to offset limited indigenous human resources. In other words, for national governments, the policy implications of immigrant labor are not confined merely to issues of control and naturalization; they involve the intricacy of how to fully utilize these endogenous resources.

Put into this perspective, the experiences of Shanghai and Hong Kong, two Chinese cities confronted by a massive influx of immigrants, reveal contrasting trajectories in policy making. There are several similarities between these two cities: First, as metropolises in coastal China they naturally attract a steady inflow of personnel looking for better economic prospects. While Shanghai is one of the most popular destinations for job seekers from rural and inland China, Hong Kong is particularly appealing to mainlanders from southern provinces, especially Guangdong. Second, immigration has become the dominant source of population growth in these cities. In Shanghai, while natural birth rate has shown a negative growth between 1995–1999, immigration has increased the population by an average of 60,000 per annum during the same period (Wan, 2000, pp. 22–24). Similarly, in the case of Hong Kong, the inflow of immigrants has overtaken natural increase as the main source for population growth. In 1996, legal immigrants from China numbered 61,000, whereas natural increase was only about 32,000 (Lam and Liu, 1998, p. 64). Third, both cities are engines of economic growth in their respective regions and are thus, for political and economic reasons, highly motivated to maintain their status at the forefront. While the economic vitality and phenomenal growth of Hong Kong hardly need further elaboration, Shanghai's status as the leading financial center of China is equally indisputable. Therefore, faced by the challenge of massive immigration, both Hong Kong and Shanghai are determined to prevent the fruits of economic progress being dissipated by an unchecked inflow of outsiders; they are, however, equally anxious to explore the full potential of extra human resources in order to maintain their global competitiveness.

Any similarities between the two cities, however, must end there. Both local governments perform contrasting roles in regulating immigration, and this leads to clear differences in the ability to utilize the contribution of immigrants. Ironically, despite the proud association of the Hong Kong government with the economic principle of laissez-faire, it is the socialist administration in Shanghai that exhibits a greater flexibility in control, allowing market forces to regulate the mix and intensity of inflow of the migrants. In the following discussion, we will see that labor mobility in Shanghai is more a phenomenon of market fluctuation and, hence, more compatible with the needs of the local economy. Prejudice and discrimination by bureaucratic departments may add extra premium for immigrants living in the city, but whether one will stay or go is very much a result of an individual's own calculation of the socioeconomic cost and benefits.

Local state policy in Hong Kong is remarkably different. Historical legacy has possibly constrained the scope of freedom and initiative enjoyed by the HK SAR government, and the consequent mind-set toward immigrants is one of restriction and exclusion. How to utilize the economic contribution of immigrants does not occupy a dominant position in government agenda. This may lead to a vicious circle and a more proactive state role in realizing that the potential of immigrant labor is warranted.

Shanghai: Bureaucratic Ambivalence and Market Logic

Defining Immigration in Shanghai

Strictly speaking, it is incorrect to define the phenomenon of this influx of outsiders into Shanghai as an immigration issue. Immigration in the ordinary sense entails change of nationality, and an engineer from Beijing or farmer from Hunan working in Shanghai obviously does not require such a change in citizenship. The residential permit system (*hukou*) in China, however, makes the situation very similar to standard regulatory practice for international mobility. Formally introduced in 1955, *hukou* has been effective in minimizing unplanned and unsanctioned movements of Chinese people across the country, particularly during the pre-reform years. Under this system, official approval is needed for residence in a place other than the locality specified in one's *hukou*; a temporary residence permit or transfer of *hukou* is required for anyone who wants to relocate. But its restriction goes beyond the freedom of movement—*hukou* basically defines one's social status, or even entire livelihood. As Solinger puts it:

> Not only the quality of the goods supplied, but the transportation conditions, the range of cultural entertainment, the nature of education offered, and the type of health care one received depended upon where one resided. And it was not just these tangible goods that were allocated by the rank of the residents, but the wages people were paid, the prices they were charged, and the subsidies and welfare benefits they received were so fixed as well. (Solinger, 1999, p. 36)

Consequently, the administration of population movement in China is analogous to the general practice of international immigration regulation. Chinese citizens residing in a "foreign" city in China have to gain the official approval of the host

administration—either a permanent transfer of *hukou* (comparable to the naturalization process) or a temporary residential permit (comparable to visa applications).
Otherwise, aliens will be treated like "illegal immigrants," vulnerable to persecution
and repatriation and living without entitlement to welfare benefits or simply at the
mercy of the host administration.

Effective Barricading of Shanghai in the Pre-Reform Years

The rural-urban segregation and restricted mobility of the population was intended
to barricade cities against rural invasion. For the party leadership in the early years
of the PRC, cities were the bastions of rapid industrialization and the root of political authority. Major cities such as Shanghai were to be protected from peasants
or job seekers from other localities. Freedom of movement had been enshrined as
a fundamental right of Chinese citizens in the first constitution of the PRC promulgated in 1954, and the inflow of population into Shanghai reflected this reality.
Between 1951 and 1954, the inflow of outsiders was 2,380,000 persons, with the
net growth of population in Shanghai attributed to population movement reaching
910,000. However, the consolidation of the *hukou* system and the municipal government's effort in mass deportation in the 1950s and 1960s reversed the trend. In
1955, 590,000 emigrated from Shanghai voluntarily or involuntarily (White, 1996,
p. 423) and the number of temporary residents also dropped sharply to the level
of 80,000 (Zhu, 2000).

The effectiveness of these measures, like other policy instruments in Mao's
China, was certainly affected by larger political and economic currents. As noted
by White:

> [E]conomic depression and political turmoil in the 1960s tended to overwhelm
> both birth control and migration control. In periods such as 1959–61, when good
> was scarce and new jobs even more so, population control was officially unin-
> tended more than intended. When the economy revived by 1963–1964, rustifi-
> cation campaigns resumed under military auspices. Urban residence control
> through neighborhood committees became more effective. By the middle of the
> decade, administrative controls collapsed, along with the administrators, into the
> Cultural Revolution. (White, 1996, pp. 425–26)

Nevertheless, the barricades against immigrants remained impermeable in general. The number of temporary residents continued to be well regulated and was
restricted to around 200,000 a year. Most of these were temporary residents who
came to visit relatives or close kin, or came to the urban area for better medical
services, staying only for a short duration. As shown in Table 8.1, the level of
inflow compared with the total population in Shanghai remained small during the
pre-reform years. In 1977, the State Council, in expectation of massive returnees
from the countryside following the conclusion of the cultural revolution, specif-

Table 8.1

Number of Outsiders as Percentage of Total Number of Residents in Shanghai City, 1955–1981

Year	Number of outsiders (O) (10,000 persons)	Number of residents (R) (10,000 persons)	(O)/(R)(%)
1955	8.01	523.74	1.53
1957	37.12	609.83	6.09
1960	26.58	641.83	4.15
1965	9.36	643.07	1.46
1971	21.08	570.74	3.69
1975	19.01	557.05	3.41
1979	17.86	591.45	3.02
1980	21.53	601.29	3.58
1981	25.93	613.39	4.23

Source: Wang, 1995, p. 29.

ically reiterated that not only was movement of people from other cities into Beijing, Shanghai, and Tianjin to be restricted, but also the transfer from the suburban and rural parts of these cities had to be controlled (White, 1996). In short, until the advent of economic reforms, the population flow into Shanghai was well policed.

Waning of Administrative Control in the Reform Period

The situation has changed dramatically in the reform era. Shanghai has become one of the most popular destinations for job seekers from all over the country and has been steadily hit by a massive influx of outsiders. Table 8.2 provides a good sketch of the intensity of the inflow.

As one would expect, most of these new outsiders settled down in the urban area of the city. By 1997, six out of seven resided in the city area (White, 1996). Even more alarming for Shanghai officials is the extended stay of these aliens. Whereas in the 1960s and 1970s most people came to Shanghai only for a short duration, "visitors" in recent years have been staying longer. A survey shows that in 1993 only about 30 percent of visitors remained a year or longer. By 1997, the ratio had gone beyond 50 percent (Wang, 1995, p. 27). In other words, Shanghai is facing a de facto immigration problem; most of these aliens are planning to settle in Shanghai (for a few years) with or without official sanction.

The majority of immigrants are motivated by the search for jobs and better economic prospects. The findings of the 1997 Population Census in Shanghai con-

Table 8.2

Inflow of Population in Shanghai, 1988–1997

Year	Number of persons moving into Shanghai
1988	1,250,000
1993	2,810,000
1997	2,760,000

Source: Zhang et al., 1998, p. 25.

firm this view with the major activities of 74.5 percent of the immigrants classified as predominantly economic in nature. The same source also shows that 69.5 percent of these immigrants had previously worked in the rural sector (Zhang et al., 1998, p. 27). The urban-rural income disparity is thus a major factor for population inflow. This is understandable; per capita income in urban areas remains more than double that of the rural area in China (China Statistical Bureau, 1998, p. 325). The introduction of the household responsibility system also makes it more sensible for rural households to find outlets for the redundant labor force at home.

Outsiders are, however, easy scapegoats for social problems (Simon, 1989). French president Jacques Chirac has stated, "If there were fewer immigrants, there would be less unemployment, fewer tensions in certain towns and neighborhoods, a lower social cost" (p. 208). Urbanites in Shanghai, who already feel endangered by the threat of market reforms, feel similar prejudice against the incomers. The number of cases of *xiagang*—de facto layoff from active employment—totaled 1,000,000 in Shanghai during the period 1992–1996 (Sun et al., 1998, p. 65). An intensification of welfare reforms leading to further reductions in subsidies and greater vulnerability to naked market forces also added little comfort to the collective psychology of the Shanghai people. A survey conducted in the new district of Pudong in the mid-1990s revealed that residents commonly believed that outsiders had brought negative influences into the city. Some 97 percent of respondents thought these incomers should be held responsible for the rising crime rate, crowded transportation, unemployment, and deteriorating public hygiene (quoted in Solinger, 1999, p. 101). One researcher has depicted an immigrant enclave in suburban Shanghai in the following way that may also reflect the general opinion on the outsiders:

> The formation of an enclave of a large number of floaters has caused a lot of social and management problems with no easy solution. The hygienic condition of the enclave is horrible, there are piles of rubbish and the environment is filthy. This is a breeding ground for epidemic diseases. The place is also a bit rough and has been a shelter and breathing space for criminals evading law enforcement.

According to the statistics of the Shanghai Public Security Bureau, there were 40,670 criminal offences in 1997, 26,074 of these cases happened in ten suburban areas heavily populated by immigrants, i.e. 64.1% of the total. And in terms of serious cases, offences committed in these areas contributed 73.9% of the total. (Sun, 1999, p. 26)

Whether the outsiders should take the blame for all the social dislocations mentioned earlier is not the issue here. Social services and facilities may have been grossly inadequate even before the massive arrival of immigrants, and the crime rate has risen in most Chinese cities during the reform period (*Ming Pao*, 11 April 2001). It is, however, this general perception, grounded or ungrounded, that sustains the prejudice. Although this may be true for any city hit by high immigration, Solinger is right in highlighting the socialist past as central to understanding the prejudice in a Chinese city such as Shanghai. Rapid population mobility serves as a painful reminder to urbanites and bureaucrats of the fading of the cherished norms of the "good old days." Immigration of the prevailing intensity is unimaginable without market reforms, but the market disrupts to the old, trusted patterns of privileges and security that define urban citizenship in cities such as Shanghai. The arrival of outsiders also exposes a confrontation between a static society governed by planning and the dynamic principle of market operation. This is a direct challenge to the bureaucrats' mind-set. Solinger has provided the following succinct description of this mentality in the pre-reform era: "At all levels in all localities, administrative management agencies plan their work and projects in accord with the size of the registered permanent population within their respective jurisdiction" (Solinger, 1999, p. 108).

The fluidity and unpredictability inherent in labor mobility is bound to be in conflict with the regimental psychology of socialist bureaucrats.

The Market Imperative

For cities like Shanghai, however, which are undergoing transition from a state-dominated economy to a proto-market, the availability of marginalized labor does provide an obvious opportunity to increase competitiveness. Flexibility in employment and a lower labor cost are among the major economic reasons for tolerating immigration. Traditional employment practice in China is characterized by secured tenure and extensive welfare; the high cost of production and inflexibility of labor management is inevitable. Workers from outside, however, are prepared to receive less-favorable terms. Those recruited through officially sanctioned channels are likely to be granted contracts with fixed tenure and minimal welfare benefits, whereas those who work illegally are willing to accept even more demeaning offers. In most cases, no insurance has to be paid, and employing young single workers also means less drainage on public finance and welfare facilities. Market forces unleashed by decades of economic reforms have obliged enterprises in state and non-state sectors to seek greater responsiveness to market changes. Central to this is the ability to react to price changes and profit margins, and swiftness in restructuring production accordingly. A floating population embodied with the desired flexibility is an option for managers preoccupied with competition and survival (White, 1994, pp. 63–94). Yet these attributes also make immigrants highly vulnerable to market fluctuations. In the face of changing demands, immigrants have to either increase their own market competitiveness or accept a lower wage. Those who fail to make either of these adjustments are vulnerable to the threat of unemployment and the growing likelihood of such occurrences certainly provokes a reconsideration of one's decision to immigrate.

A careful analysis of the composition of Shanghai immigrants confirms that market forces are a dominant factor regulating influx intensity. As shown in Table 8.3, while the size of immigration doubled between 1988 and 1993, there was a slight decrease of 50,000 between 1993 and 1997. In recent years, the inflow has been discouraged by several developments in Shanghai. First, with most works of urban redevelopment accomplished by the early 1990s, the growth of the Shanghai property market has remained sluggish. Consequently the growth of the construction industry—one of the major sectors absorbing immigrant workers—has been affected. The intensification of employment reform in SOEs also implies an increase in the supply of labor and hence makes job seeking a more daunting task. Demand for labor, in other words, has decreased during this period, and this has put an effective brake on the inflow of population into Shanghai.

The intensity of competition also affects the mix of immigrants. The general educational level has increased, with workers with junior high education or above constituting almost three-quarters of the total immigrant force in 1997, whereas less than 60 percent had attained these levels in 1988. Young people from the age group of twenty to forty years also commanded a larger proportion of the total labor force in 1997; 60.9 percent fell into this range in 1997 compared with 51.1 percent in 1988. In short, the changing market situation has automatically exerted a selection effect on immigration: with survival becoming more difficult, the younger and better educated are more likely than others to take up the challenge of moving into Shanghai.

The market selection effect is also confirmed by the findings of a survey on immigrants' income. A study conducted between 1995 and 1997 shows that the income of immigrants has been rising; the average monthly income jumped from 611 to 780 yuan between 1995 and 1997 and the gap between immigrants' income and Shanghai residents' has also been narrowing (Liu and Shang, 1998). In other words, their employment may no longer be confined to the low-paying physical work of the immigrant stereotype; they are also sought in posts that require greater sophistication and earn greater reward.

Government Role in Utilizing the Market Force

There are two major components to the general strategy of the Shanghai government toward population influx: containing the number and minimizing its drainage

Table 8.3

Background of Immigrants into Shanghai City, 1988–1997

	1988	1993	1997
Number of inflow (10,000 persons)	125	281	276
of which come from outside Shanghai	106	251	237
Age composition			
Below 20	25.7	20.8	19.7
20–29	32.6	41.7	39.2
30–39	18.5	19.7	21.7
40–59	15.7	14.3	15.7
60 or above	7.5	3.5	3.8
Educational background (%)			
Tertiary education or above	1.9	2.0	4.7 (11.5)
Senior high	11.2	10.1	12.7 (26.3)
Junior high	45.3	50.9	50.2 (36.7)
Primary school	27.8	28.0	21.8 (18.1)
Illiterate or semi-illiterate	13.7	9.1	6.0 (7.8)

Source: Zhu, 2000, p. 3; China's Statistical Bureau, 2000, pp. 104–5.

Figures in ( )—percentage of Shanghai population in 1999.

on public finance. As long as the size of inflow is confined to a socially and financially acceptable level, the Shanghai government is willing to act flexibly. Intentionally or unintentionally, this pragmatic regulatory regime facilitates market forces by orchestrating the movement of the floating population in harmony with the economic needs of Shanghai.

According to the Five-Year Plan on the Floating Population, the primary objective of the Shanghai government is to control the size of population inflow below the level of 3.5 to 4 million. Of this population, the number of those who stay in the city for six months or more should be kept below 1.5 to 2 million. The latter are, in fact, de facto immigrants and are the main concern of the government on this policy issue.

There exist differentials in official treatment toward these 1.5–2 million "visitors" of extended stay:

1. Granting of temporary residential permit (*zanzhuzheng*). An individual from another locality residing in Shanghai city can obtain a permit from the Shanghai public security bureau if he or she has a valid Chinese citizen identity card, proof of residence, and proof of employment or steady income. This document is valid for two years and exempts the holder from repatriation (Shanghai People's Congress Standing Committee, 1996).

2. Granting of a temporary work residential permit (*gongzuo jizhuzheng*). Unlike the temporary residential permit, this document is only obtainable through an employer's application to the municipal government Bureau of Personnel. The scheme is mainly confined to professional or management staff employed in Shanghai. A stricter reviewing process is stipulated and a holder who has worked the same work unit for three years or more is entitled to apply for formal transfer of *hukou* into Shanghai (Shanghai Personnel Bureau, 1998).

3. Acquiring a blue-chop *hukou* (*lanyin hukou*). This requires meeting one of these conditions: (a) for foreigners and investors in Hong Kong, Macau, or Taiwan, making an investment in Shanghai of at least U.S.$200,000, or 1,000,000 RMB yuan for Chinese nationals of more than two years; (b) buying a foreign-sale commodity flat (*waixiao shangpinfang*) with a construction area above 100 square meters; and (c) by possessing professional skills and making contributions to their enterprises. The blue-chop *hukou* is still classified as a temporary registration, but the holders are entitled to certain benefits previously reserved for permanent residents (Wong and Huen, 1998).

4. Permanent transfer of *hukou* into Shanghai (*qianru hukou*). This means the termination of alien status, making the holder a legal resident with full entitlement to the welfare and civil rights bestowed on any permanent *hukou* holder in Shanghai.

Clearly, a gradient of entitlements is evident in Shanghai's regulation of immigration. Individual economic contribution and degree of self-sufficiency are the factors determining one's position in this hierarchy. To local officials, those who are less competitive, have difficulty in finding jobs, and have no proper shelter are likely to be a drain on local resources and therefore unwelcome. Regarded as parasitic, and at the root of social problems, they will be subject to repatriation. Yet those who are lucky enough to find themselves steady income and accommodation are granted a wide spectrum of permits sanctioning their residence in Shanghai. The treatment varies according to one's income, professional skills, and employer's recommendation. The greater the contribution, the greater the security of residence. Permanent transfer of *hukou* is the ultimate prize for those who have their potential recognized by local government. The system reinforces the market selection effect mentioned earlier. Administrative hurdles help to drive away the least competitive, but the system also rewards the more competent. In this way, the social cost and financial resources needed for supporting the immigrants are minimized, and the potential of outside talent is fully exploited.

An analysis of the educational profiles of those transferred *hukou* into Shanghai confirms this view. If we restrict the analysis to those who have permanently transferred their *hukou* to Shanghai, the educational attainment is highly impressive: 25.1 percent of employees with transferred *hukou* reached tertiary education or

Table 8.4

Occupation Composition of Shanghai Immigrants (%)

Occupation	Total labor force	Immigrant labor with transferred *hukou*	Immigrant labor without transferred *hukou*
Professional/ technician	14.9	24.7	4.2
Leading persons in bureaucratic unit, enterprise, or party-mass organization	5.9	6.9	3.4
Supporting staff in these units	7.9	11.4	3.2
Commercial sector	8.8	9.7	14.6
Service sector	8.5	13.4	9.7
Agriculture, fishery, and husbandry	11.4	2.7	10.2
Transportation and manufacturing sector	42.3	31.2	54.7
Others	0.3	—	—

Source: Gao, 1997, p. 47.

above, whereas only 11.4 percent of Shanghai's total labor force have such qualifications. The average length of education of the former is also higher, that is, 11.5 years compared with 9.9 years (Gao, 1997).

Further evidence of this trend can be seen in Table 8.4. About a quarter of the labor force with transferred *hukou* fall into the professional or technician ranks, whereas only 4.2 percent of immigrants without permanent *hukou* transfer work at this level. We also find a smaller percentage of people with transferred *hukou* involved in manual labor (31.2 percent). The proportion of other immigrants is much higher (54.7 percent). It is also notable that those with transferred *hukou* are doing better than the Shanghai population in general. In other words, they are more competitive and a possible means for improving the city's productivity.

Hong Kong: Constrained Autonomy in Regulating Immigration

Whereas one may find many similarities in the approach adopted by the Shanghai government as compared with many other cities or countries in regulating immigration, the case of Hong Kong may be unique. In this section, we concentrate mainly on the administration of immigrants from Mainland China, as this source constitutes the dominant majority of immigration to Hong Kong. What distinguishes Hong Kong from other cases is its lack of autonomy in controlling the influx of immigrants into its territory. Put simply, owing to its colonial legacy, the Hong Kong government has to share its regulatory power with Chinese authorities at central and local levels. Consequently, its ability in utilizing the economic contribution of immigrants has been constrained, and the issue is perceived more as a drain on indigenous resources rather than a potential addition of human wealth into the domestic economy.

From the Nonexistence of Border Control to the Quota System

"The history of Hong Kong is a history of Chinese immigrants," one commentator has astutely put (Lee, 1998, p. 267). Since the British takeover of Hong Kong from the Qing government in 1841, the population of the territory has increased from 2,000 to more than 6 million, of which more than 20 percent came from Mainland China in the 1960s. In other words, people residing in Hong Kong are mostly immigrants themselves or descendants of immigrants. However, the administrative framework for this steady inflow of population has undergone drastic change during this period.

1841–1950s: Absence of Border Control

The British adopted a laissez faire policy toward the inflow of population from Mainland China during this period. Mainlanders could come and go or even settle in Hong Kong freely without prior approval from the Hong Kong administration. There was a practical need for such a leniency in border control. For the British, the tiny population of the fishing village they took over from China in 1841 was insufficient to sustain its industrial and commercial development. There was, in other words, a shortage of labor, and China could provide a convenient source of supply. By 1902, the population had reached the figure of 360,000 (Yuan, 1998, p. 10). Instability on the mainland was the major factor for population influx into Hong Kong during this period. The outbreak of the Anti-Japanese War in the 1930s triggered an exodus, and in 1941, the population of Hong Kong exceeded one million for the first time. The civil war between the communists and nationalists in the late 1940s led to another explosive wave of immigration, and the population had reached the mark of 1.8 million by 1948 (Yuan, 1998, p. 10).

1950s–1980: Formalizing Administrative Control

The founding of the People's Republic of China in 1949 created a new complication for the Hong Kong administration concerning immigration control. Fear of communist rule led to a new wave of immigration into Hong Kong in the early

1950s. By mid-1950, the population of Hong Kong exceeded 2.5 million. As many scholars have rightly pointed out, this fact was crucial for the making of the economic miracle in Hong Kong, as among these immigrants were personnel with entrepreneurial skills, know-how, and capital (Wong, 1998). However, the drastic growth in population exerted enormous pressure on Hong Kong. The colonial administration was understandably anxious to halt this massive inflow. However, unlike the Qing and Republican governments, the new communist regime was much less compromising on nationalistic grounds. According to the communist leaders, the three treaties signed in the nineteenth century confirming the concession of Hong Kong to Britain were regarded as unequal and hence, illegitimate. In short, the communists would not recognize British rule over Hong Kong, though they were not prepared to use force to regain sovereignty of the territory. This stance created a problem for the British. The Chinese claimed that the colonial government in Hong Kong simply had no right to stop mainlanders entering into another Chinese locality. The British, however, responded with great pragmatism, and a compromise was reached. Under this agreement, the administration of immigration into Hong Kong was confirmed as the right of the Chinese government. They could solely decide on matters like the number of migrants and the vetting and approval of applications. Those who passed the official scrutiny of the Chinese authority would be issued a one-way permit to Hong Kong, and the colonial administration had to accept them. In return, the Chinese government promised that due respect would be given to the concerns of the Hong Kong administration when deciding the number of one-way permits. In addition, under the new arrangement those who entered Hong Kong without permits would be categorized as illegal immigrants and could be repatriated by the Hong Kong authority. This system worked well during the early years, and the net balance of departures and arrivals in Hong Kong averaged about 5,000 to 6,000 per year from 1952 to 1956 (Lam and Liu, 1998, p. 9).

1980 Onward: Tightening of Control

The larger historical changes in China and Hong Kong would soon, however, shake the foundations of the new regulatory framework. Political turmoil and economic dislocation resulting from the disasters of the Great Leap Forward, and the cultural revolution exerted a strong push for a new wave of exodus. On the other hand, both the open door policy adopted by the post-Mao leadership and the economic takeoff of Hong Kong in the 1970s led to greater awareness among mainlanders of the huge gap in living standards between Hong Kong and China. Consequently, the flow of illegal immigrants into Hong Kong increased dramatically toward the end of the 1970s. As shown in Table 8.5, there were more than 400 illegal immigrants caught at the border every day. When considered together with the 55,000-plus legal immigrants with one-way permits, it can be seen that the population

Table 8.5

Illegal Immigration from China, 1970–1990

Year	Annual	Per day
1970	3,416	9.4
1971	5,062	13.9
1972	12,958	35.5
1973	17,561	48.1
1974	19,800	54.3
1975	8,250	22.6
1976	8,054	22.1
1977	8,361	22.9
1978	19,438	53.3
1979	192,766	528.1
1980	150,089	411.2
1981	9,220	25.3
1982	11,160	30.6
1983	7,604	20.8
1984	12,743	34.9
1985	16,010	43.9
1986	20,539	56.3
1987	26,707	73.2
1988	20,808	56.9
1989	15,841	43.4
1990	27,826	76.2
1991	25,422	69.6
1992	35,645	97.9
1993	37,517	102.8
1994	31,521	86.4
1995	26,824	73.5
1996	23,180	63.5
1997	17,819	48.8
1998	14,613	40.0
1999	12,170	33.3
2000	8,476	23.2
2001	8,322	22.8

Source: 1970–1990 figures, Lam and Liu, 1998, p. 12. Figures from 1991 onward, Hong Kong Immigration Department.

pressure faced by Hong Kong in the late 1970s was alarming. The Hong Kong government responded by adopting a stricter policy toward granting residence status to illegal immigrants. In 1980, it abolished its "reached-base" policy—a practice that allowed illegal immigrants who had successfully crossed the border and entered the urban area uncaptured to remain in Hong Kong. Thereafter, all those arrested upon arriving illegally were repatriated to China. New legislation was also introduced requiring all adult residents in Hong Kong to carry identity cards at all times. In the same year, the Hong Kong and Chinese governments also agreed to restrict the quota of one-way permits to 150 a day. This was down from a peak of 310 in 1978. The daily quota was further reduced to 75 in 1983 and remained at this level until 1993 (Lam and Liu, 1998, p. 10).

Family Reunion as the Major Motivation for Immigration

The major challenge for Hong Kong immigration control, however, goes beyond the issue of restricting numbers. The removal of the right to screen prospective immigrants from China suggests the inability of the Hong Kong administration to regulate the inflow of population in accordance with its social and economic planning. What makes the situation even worse for Hong Kong is the strong pressure for family reunion. The growing social and economic exchanges between Hong Kong and the mainland since the 1980s have contributed to the rising number of cross-border marriages. According to the 1997 General Household Survey, it was estimated that 112,000 Hong Kong residents had spouses and some 162,000 children living in China (Census and Statistics Department, 1997). The number of children was estimated to be in the region of 700,000 in 1999 (Chan, 2000). While economic skills and competitiveness may be the major attributes needed to gain admission into Shanghai or other countries, immigrants into Hong Kong are mostly considered on compassionate grounds. According to the Chinese authorities, citizens are eligible to apply for one-way permits to Hong Kong if they are:

1. People with spouses living in Hong Kong and who have been apart for many years
2. People with elderly parent(s) living in Hong Kong in need of care
3. The elderly or children who have no one to depend on and want to reunite with close kin or direct family in Hong Kong
4. People who are the sole inheritors of property of deceased relatives in Hong Kong
5. People with special reasons (State Council, 1986)

Categories (1) and (3) are the priorities among all the applicants. In 1998, the Chinese government further refined the procedure by introducing a points system for applicants. Applicants of category (1) are given 0.1 point a day with the date of marriage as the starting point. Those who reach 182.5 points are given the one-way permit, meaning they should be granted admission into Hong Kong within five years. Child applicants also take up more than one-third of the daily quota of one-way permits issued every day (*Ming Pao*, 17 July 1997). The right of children of Hong Kong residents to reunite with their parents is also enshrined under article 24 of the Basic Law of the Hong Kong SAR Government—the miniconstitution of Hong Kong after 1997.[1]

The consequent demographic composition of the immigrants is alarming. According to the 1999 Census Survey, more than 35 percent of immigrants from the Mainland are outside the age group of active labor (below 15, and 60 or above); the figure in Shanghai is substantially lower (age groups of below 20, and 60 or above constituting less than 23 percent—see Table 8.3). A similar pattern can also be observed in the distribution of educational attainment of immigrants (see Table 8.6). As much as 35.4 percent of Chinese immigrants into Hong Kong have only primary education or less, whereas the figure is restricted to 27.8 percent in Shanghai (see Table 8.3). Consequently, a large proportion of these immigrants remains economically inactive, and this constitutes a heavy drain on resources. According to the 1999 Census Survey, the percentage is as high as 57.3 percent. Again, Shanghai enjoys a more favorable rate in this aspect. Only 24.8 percent of the immigrants there were not involved in production in 1997 (Zhang et al., 1998, p. 68). The fact that the percentage of economically inactive immigrants has been growing since 1981 gives little comfort to the Hong Kong government (Lam and Liu, 1998, p. 50). The way the Chinese authority issues one-way permits is also subject to local criticism. Applications from individual members of the same family are not considered together, and thus they may be granted permits at different times. Consequently, many single-parent immigrant families result. Not only does this contribute to a low labor-participation rate among immigrants, but it also increases the demand for welfare support. Without official control of immigration, the government remains unable to regulate inflow according to the economic needs and developmental priorities of the territory. Also, with family reunion the prime concern dominating the approval process of one-way permits, the effectiveness of the market mechanism in selecting the more competent for the domestic economy is diluted.

Most unfortunate for the Hong Kong government is the preservation of the status quo by the Chinese government in spite of its resumption of sovereignty. Although the proclaimed rationale of the Chinese authority's monopoly of migration control was the nonrecognition of the British legal position on Hong Kong, there was little justifiable ground for Beijing to observe the same stance after 30 June 1997. Hong Kong is no longer a British colony, but unfortunately Beijing has shown no sign of surrendering its monopoly over this matter. Article 22 of the Basic Law reiterates the passive role of the Hong Kong administration:

> For entry into the Hong Kong Special Administrative Region, people from other parts of China must apply for approval. Among them, the number of persons who enter the Region for the purpose of settlement shall be *determined by the competent authorities of the Central People's Government* after consulting the government of the Region [emphasis added].

This unreasonable policy has aroused growing discontent in Hong Kong (*Tin Tin Daily*, 16 April 2000). Twenty-two delegates from Hong Kong in the National People's Congress (NPC) issued a joint statement on this matter to the Central

Table 8.6

Social Characteristics of Chinese Immigrants, 1999

	Percentage
Age group	
0–9	15.9 (11.03)
10–19	27.1 (13.59)
20–29	11.0 (15.23)
30–39	21.7 (19.61)
40–49	16.2 (17.26)
50–59	5.2 (8.51)
> 60	2.9 (14.76)
Educational attainment	
No schooling/kindergarten	6.9 (8.4)
Primary	28.5 (22.3)
Secondary/matriculation	58.8 (51.7)
Non-degree course	2.7 (6.9)
Degree course	3.1 (10.7)

Source: Hong Kong Census and Statistics Department.

Figures in ()—percentage of Hong Kong natives.

People's Government in 1999. They tried in vain to demand a greater role for the HKSARG in administering the one-way permit system (*Ming Pao*, 15 March 1999).

Half-Heartedness in Tapping Talent from China

The Hong Kong government is not unaware of the importance of tapping talent from abroad. In fact, as seen from Table 8.7, more than 14,000 work permits are granted to expatriates every year. There is no quota or restriction on the number of work permits issued each year, and the Immigration Department evaluates each case on the basis of the merits of the applicant and the needs of the domestic economy.

The procedure for recruiting professionals from the mainland, however, is more stringent. In the late 1990s, the government introduced a number of schemes targeting mainlanders. The desire to tap the skills of talented people from China was certainly a motivation. However, even more pressing may have been the anxiety of the local business sector to reduce production costs. The shock of the financial crisis in 1998 aroused great concern among businessmen and officials about the competitiveness of Hong Kong; local productive forces were diagnosed as overpaid and underskilled. The growing economic integration between Hong Kong and the Mainland further necessitates the government to reconsider its policy on recruitment

Table 8.7

Distribution of Work Permits, 1998–2000

Industry	1998	1999	2000
Merchant banks	238	257	219
Technical professionals	2,891	2,478	4,187
Executive, management, & other professionals	7,373	6,947	8,297
Others	3,885	4,080	5,209
Total	14,387	13,762	17,912

Source: Hong Kong Economic Journal, 13 March 2001.

from the latter. There is a genuine risk that if local employers cannot tap cheap Mainland talent, many businesses will simply relocate north of the border. Consequently two new schemes have been introduced since 1999.

The Admission of Talents Scheme 1999

The scheme sets no limit on quota, industry, or wage level for successful applicants. The admitted can take their spouses and family to reside in Hong Kong, and the right of abode will also be granted to them after seven years of continued residence. However, the scheme is restricted to top-level professionals. Only eligible are people with doctorate degrees obtained in the top two hundred universities in China or with proven success in research or professions. The demanding qualification requirement has deterred interest from local businesses. Since its implementation in December 1999, there have been only 452 applications with only 117 approved (*Hong Kong Economic Journal,* 13 March 2001). The scheme clearly does not address the genuine concerns of local economy.

The Revised Admission of Professionals Scheme 2001

Unlike the above plan (which is aimed at the professional elite engaged in innovation and research), the revised Admission of Professionals Scheme is intended to meet employers' immediate operational needs. The qualification requirement for applicants is lax, as proof of technical knowledge, rather than a bachelor degree is requisite. The scheme aims to provide an opportunity for local businessmen to employ midlevel technical or executive staff from the mainland. As Financial Secretary Donald Tsang put it:

> The admission of professionals will bring about a much-needed transfer of knowledge and experience from the mainland. . . . This will be invaluable for us in the

competition for an expanded China market. Failure to take immediate measures, when other economies are already making similar moves, will undermine Hong Kong competitiveness. (*South China Morning Post*, 8 March 2001)

The new scheme also provides greater employment flexibility for mainlanders studying in Hong Kong. They are allowed to stay on for employment after graduation provided that they meet the eligibility requirements. No quota has been set for the scheme. Nonetheless, the scheme only applies to two sectors: information technology and financial services—areas the government believes are facing the most severe shortage of qualified personnel.

These two schemes are obviously steps forward that enable the local economy to attract talent from the mainland. Nevertheless, the cautionary stance of the Hong Kong government is still evident. A double standard in handling the admission of professionals from the Mainland and other countries is clear, and discrimination against the former is apparent. While expatriates from other countries face no specific restriction on industry, quota, or academic qualification, the eligibility requirements for the mainlanders are extensive, in spite of the changes mentioned earlier. The government's fear of the massive inflow of professionals might be justified given the substantial discrepancy in wage levels and living standards between Hong Kong and the Mainland. However, the strong phobia against mainlanders in local society may be an even more compelling explanation for the hesitation and indecisiveness of government policy. It may seem surprising to find such a common discriminatory stance toward mainlanders in a society where most residents are themselves second- or third-generation immigrants, and also of the same racial stock as the mainlanders. However, such popular resentment against Chinese immigrants is undeniable. An opinion poll conducted in 1998 showed that a higher percentage of respondents (28.2 percent) had a negative impression of mainlanders than positive (18.2 percent) (Social Science Research Centre, 1998a). Another focus group study conducted by the same institute reveals that most participants perceived mainlanders as "poorly educated, autocratic and corrupt" (Social Science Research Center, 1998b). Decades of colonial rule, the memory of the June 4 tragedy, and the parochial outlook of the local population go some way to explain the popular hostility and arrogance toward mainlanders. Yet the HK SAR government also plays a part in manufacturing this negative sentiment. In 1999 in order to win popular support for its challenge against the ruling of the Court of Final Appeal on the right of abode for immigrants' children, the government deliberately stirred up fear and reinforced hatred against mainlanders. Its overall strategy in this episode was to portray these potential immigrants as parasites on Hong Kong's prosperity. The potential impact of prospective immigrants on the local economy was also inflated by government estimates. According to its analysis, taxpayers would face a bill of $28 billion to provide new schools, and the unemployment rate would rise by 25 percent. The standard of social services would also inevitably be affected, with the health service returning to the bad old days of crowded hospitals with

Table 8.8

Summary of the Challenges and Responses to Immigration of Hong Kong and Shanghai

	Shanghai	Hong Kong
Intensity of the inflow	In the region of 2.5 million by 2000	Annual intake of 55,000 legal immigrants and about 10,000 illegal immigrants
Attributes of immigrants	80% within active labor age 70% with junior high education or above	65% within active labor age 65% with junior high education or above
Major reason for immigration	Economic	Family reunion
Major means of administrative control	Residential permit system	One-way permit system
Effectiveness	Impose higher cost of living for violator, but its general effect of deterring unauthorized residence weakened by the advent of economic reforms	Reinforced by stringent border control of both Hong Kong and Chinese forces, effective in checking illegal immigration Screening and approval of immigration application in the hands of Chinese authority
Schemes to attract talented immigrants	Flexible and accommodating	Cautious and rigid

camp beds. The accuracy of the government's estimate is highly debatable, as has been pointed out by many academics and politicians (Chan, 2000). However, its strategy of arousing fear of immigrants appeared to be hugely successful. The debate on the schemes to admit Mainland professionals is revealing. The Democratic Party, the largest political party in Hong Kong, had taken a spirited stance in defending immigrants' right of abode. However, it is regrettable to see that even this party has succumbed to the populist fears and advocated blocking the move to admit more Mainland professionals (*Ming Pao*, 16 March 2001; *Hong Kong Economic Journal*, 12 March 2001). This is definitely a clear sign of the prevalent phobia and resentment against mainlanders.

Comparative Overview

In short, the major challenges and responses adopted by Hong Kong and Shanghai toward the issue of immigration are summarized in Table 8.8.

Conclusion

> Hong Kong and Shanghai are like a pair of strikers in a football team.
> Competition exists between the partners, but whoever scores benefits the
> other too.
>
> —Xu Kuangdi, the mayor of Shanghai

China has made an indisputable contribution to the Hong Kong miracle. China has always been for Hong Kong the source of cheap food, water, electricity, and other daily subsistence for average households, and has hence helped to maintain production costs at a low level. Economic reforms in the post-Mao era also provide adventurous businessmen in Hong Kong with lucrative markets and opportunities. However, the close link is a double-edged sword. Exchanges are not confined to the interflow of capital, goods, and experiences; communication between people is also on the rise. Growing human exchanges bring new vision and skills necessary for further prosperity, but these trends also breed more cross-border marriages, as well as legal and illegal immigration. Hong Kong has its historical luggage in handling immigration, but it is not alone in facing the challenge of a massive influx of population from a less-developed proximity. As seen from the preceding analysis, Shanghai faces no easier a task. However, what distinguishes Hong Kong from cosmopolitan cities like Shanghai is its failure to thrive on an open immigration policy that would allow it to utilize the skills of talented people from other places, particularly the Mainland. The local authority in Hong Kong is inclined to restrict the human inflow from the Mainland indiscriminately. Skilled or unskilled immigrants are in general treated with caution, if not contempt. The colonial legacy and the irrational fear among the local population have obstructed the convergence of the development needs of the market and the abundant supply of skills and experience possessed by prospective immigrants north of the border. The authoritarian character of the socialist administration in Shanghai has ironically enabled the city to be more pragmatic in handling immigration. "Restricting inflow of population, not talent" (*xianzhi renkou, bu xianzhi rencai*) is the proclaimed guiding philosophy of the population policy. While the HK SAR government remains obsessed with its tradition of passive restriction, the Shanghai government continues to act as an enabling force for market selection of talent. The administrative framework provides different levels of recognition and privileges for immigrants, which vary according to one's marketability. It closes the door for the less competitive, and hence parasitic, but it endorses and rewards the survivors and the best. The administrative ethos is, in other words, very much in line with the jungle rule of market economy. For Hong Kong, immigration will continue to be regarded as a drain on local prosperity, but for Shanghai, there seems to exist a genuine possibility that among these ugly ducklings at least a handful will turn into graceful swans (bringing glory and charm to the city).

Mayor Xu is right in pointing out the complementary nature between Hong Kong and other Mainland cities, such as Shanghai. While the Hong Kong business and professional sectors have enthusiastically embraced the business opportunities and initiatives of their Mainland counterparts, the city's conservative population policy can hardly put it on a equal footing with Shanghai in the scramble for talent from other Mainland cities. This constitutes an obstacle for Hong Kong in maintaining competitiveness in the global market.

Note

1. A major controversy over the issue of the right of abode for children of Hong Kong residents broke out in 1999. The Hong Kong Court of Final Appeal ruled that children born to permanent residents had the right of abode in the HK SAR, whether their parents were permanent residents of Hong Kong or not at the time of their birth. This caused a constitutional and political crisis in Hong Kong and the verdict was finally reversed by the intervention of the National People's Congress of the PRC.

Facilitating Fortunes vs. Protecting People in China's Richest Cities

Lynn White

This tale of two cities is also a story of antinomies. Hong Kong and Shanghai are the most modern cities in the political system with the world's oldest continuous heritage. These are metropolises with "first world" economies, but their recent prosperity depends on quick development of the neighboring Pearl and Yangzi River Deltas, where labor is hired at "third world" wages. Each of their social policies reflects this situation.

Their age structures include many elderly people who need health care, as well as young students from ambitious families who demand high-quality education. There is no full agreement in either city on who will pay (and how much) for costly modern medicines, elder care, or schools. Housing is often expensive and oddly priced in both cities. In both, competition for residential and commercial space has bred unexpected kinds of urban politics. Social welfare is not a project that either municipal government is prepared to finance endlessly. The border of the Hong Kong Special Administrative Region, like the household registration system in Shanghai, is designed to reduce in-migration, regulating unskilled labor. Yet modernization in these cities also demands heavy construction work that middle- and high-income people refuse to do, so the immigration of poor people is both permitted and disdained. Within each city, this book finds two tendencies in tension with each other: to raise capital from society and to avoid spending it on social support.

Differences between Hong Kong and Shanghai social policies are detailed in tables in the introductory and functional chapters, and further broad differences can be summarized here. Shanghai has twice as many people as Hong Kong. Its terrain is sedimentary and flat, not basaltic and mountainous like Hong Kong's topography. So Shanghai has space for heavier industries. The former Crown Colony is much richer on a per capita basis than China's most productive, province-level "directly ruled municipality" (*zhixia shi*). Hong Kong had a GDP per person over U.S. $26,000 in 1998, higher than the rate in Britain, while the comparable figure in Shanghai was one-eighth as much. The per-capita economic gap between the two cities has lessened recently, because Shanghai's growth has been faster. Disposable personal incomes, especially when measured in terms of purchasing power, are also less different than GDP figures suggest.

Tax rates in Shanghai, especially for companies, have been much higher than in Hong Kong. During Mao's time, Shanghai was proudly recorded as the largest provincial contributor to central PRC coffers. Shanghai's phenomenal growth after 1990 depended on its surrender of that honor to other provinces—first to Jiangsu, which now borders Shanghai and of which the city was historically a part. By contrast, Hong Kong has a famously low tax regime, and unlike Shanghai it sends no formal remittances to Beijing.

Another difference is that Hong Kong's borders are more hermetic than Shanghai's. The household registrations (*hukou*) of the Maoist era have now practically dissolved into more varied gradations of urban legitimacy for urban residents. Many migrants to Shanghai have come without the benefit of permanent registration, especially during the early 1990s when the city had even more construction sites than it now does. Shanghai still tries to register its migrants who bring capital or skills. Its difference with Hong Kong is greater ease of illegal immigration among people who bring no money or lesser talents—but these workers can also raise urban efficiency. Hong Kong has a higher fence, which the Mainland government has controlled for decades. There has been some influx of migrants, although Hong Kong residents resent new arrivals more bitterly than might be expected in a city where most families arrived just one or two generations earlier. Migrants to the Special Administrative Region are legitimated when they are in temporary, excludable categories (e.g., Filipina amahs). Shanghai's more permeable border, together with the recent high-tech prosperity of several Jiangnan cities that are closely associated with Shanghai, suggests the East China metropolis is less insulated from both inland successes and inland poverty.

The great similarity of these two cities, when compared with other parts of China, is their wealth. Shanghai grew pell-mell in the 1990s, reportedly at a 13 percent annual rate from 1992 to 1998. It had lagged during the 1980s, when its light industries were restructured too slowly to remain competitive with inland firms. Hong Kong's economy grew more steadily, from a much higher base to a "first world" level before the turn of the millennium. Both urban economies experienced major realignments during the century's last two decades. For Hong Kong, the main change was a shift of capital and low-wage jobs into China, while higher-compensated work remained in Hong Kong. For Shanghai, the change was more complex, because heavy industries from Mao's time were capital-intensive, not easily moveable, and under less competitive pressure than light industries. These cities adapt to their environments. External changes affect what governments can do with inherited structures and how much money is available for social policies.

Business-Government Decision Structures

The Hong Kong government is more exclusively for, by, and of businesses than is any other regime on this planet. Modern states usually claim to be for, by, and of "the people." In Shanghai, entrepreneurs also increasingly dominate policy. Their closer connections with a very ex-proletarian party, an army, and Third World inland places nonetheless create slightly less focus on the politics of quick profits. Shanghai's government has officially defined a poverty line, and it gives some help to registered households with incomes below that level. The Hong Kong government demurs to commission a study that might determine any official poverty line, lest its costs rise for helping people who are that poor. Colonial officials' insistence on "flexible prices" ensured their minimal support for wage bargaining. Hong Kong's export of low-wage jobs has raised unemployment in the territory. This trend doubled, from 1995 to 1999, the number of people who received comprehensive Social Security Assistance. Despite official hopes of changing welfare to workfare in this environment, the government's budget for social security soared almost ten times in these years. Neither of these two city regimes hopes to squander its capital without raising its productivity, but economic downturns create structural problems for citizens.

Hong Kong and Shanghai have both experienced periods of planning for exports. This means exports not just to foreign markets. It also means exports of goods and services to other parts of China—in Shanghai's case especially before the 1980s and in Hong Kong's since 1990. Both cities have also experienced periods of planning for educational and technical upgrading in the labor force. These municipal governments have wanted the local economies to expand into new technologies and raise average productivity levels.

Changes of plan have come in Hong Kong, not just Shanghai. Both urban regimes strongly tend to prefer market solutions over state solutions to all social policy problems. The viability of policy choice for either the market or state style, however, depends on money, economic vibrancy, the tax base, and the ability of the government to regulate the economic sectors that are most quickly expanding. China's socialism became less feasible when its very autonomous state was faced with burgeoning rural and suburban industries that undermined remittances to the public treasury. Hong Kong's largely unadmitted planning, which will be described more fully, ran into trouble when the territory's historically special place in China's economy was eroded by the opening of new ports and other successes elsewhere in the nation. This trend reduced the unique rent that the city's main governors (who are in business) had been able to extract, and it literally lowered the value of Hong Kong as land. These changes severely strained the resources available for social welfare.

In Shanghai, the demise of planning was fastest in the mid-1980s. Rural and suburban industries by then had captured both raw materials and product markets that had previously sustained the mostly state-owned enterprises (SOEs).[1] These had once been subject to enforceable plans, and their profits supplied most of the socialist government's budget. Shanghai not only had been the foremost contributor to China's national treasury, it had also sustained decent support for local workers whose residence registrations the police could still monitor. Inland booms brought an end to this social order, both by reducing the portion of resources the central authorities controlled and by creating a more mobile labor market and urban im-migration.

In Hong Kong, Chinese business elites and expatriate colonials alike have imagined an almost totally laissez-faire, nonstate regime. Gurus in the Heritage Foundation still declare Hong Kong to be the world's most free economy.[2] Imposts charged by government (taxes) are indeed low and flat. For many decades before the mid-1950s, the British regime minimized all official costs (except for law and order). Social services, when available at all, were provided by philanthropies, as well as religious or lineage organizations.[3] All regimes tend to develop stubborn "path-dependent" styles of rule, and Hong Kong's style dictated that most money should stay in businesses and the government should spend little.[4] This was not, however, a lack of planning. It was a *type* of planning by local elites, who historically remained outside the colonial government although some of them were (and are) at least as powerful as the top officials.

This pattern was long ago challenged practically, but not reversed ideologically, by a huge fire among squatter huts at the Shek Kip Mei tract, northeast of Kowloon, in late December 1953. This event came after Labour Party governments in Britain had initiated a spate of welfare programs at a time when colonies were ending in many parts of the world. The city's late-imperial leaders then faced a communist regime at the peak of its patriotic popularity on the mainland. Breaking previous patterns, the government decided to spend some money. It began to build and manage housing estates for unskilled laborers.

Hong Kong thus developed an atypical form of economic planning. It did not try to pick specific leading industries, head geese as in the "flying goose" model that Japan perfected for both import substitution and export promotion.[5] Instead, the government at this stage tried to aid the competitiveness of *all* Hong Kong firms by means of indirect subsidies to wage bills. Companies had to cover (through salaries) only the low, fixed rents for the public quarters where their workers lived. Especially after the influx of mainland refugees from the early 1960s famine in China, a rising portion of all Hong Kong's people lived in government flats. A few years later, half did so. Most low-income families rented, rather than purchased, their state-built homes. The state subsidized their housing.

The government also controlled the land price, keeping it high and carefully monopolizing sales of new property as a means to raise revenues. This plan allowed Hong Kong to retain its tradition of low direct taxes on corporations and individuals, while garnering large amounts of public money for social services. Land-sales profits and other property taxes accounted, at the high point, for as much as 54 percent of the government's total revenues.[6] Selling land allowed the state to continue its generalized subsidy to businesses' wage bills and meet increasing educational and other expenses, while encouraging private philanthropies (which in that era could not easily work elsewhere in China) to serve the needs of the people also. Aside from low-cost housing, somewhat less-massive indirect subsidies in Hong Kong's planned low-wage regime went to water and food. Calculation of the total extent of this money shows that, at least by the start of the 1980s, half the cost of labor provided by private companies' employees who lived in public housing was, in effect, indirectly paid by the government.[7]

Was this a pure market economy? Liberal economists such as Milton Friedman still believe that Hong Kong meets their theoretical ideal: a market without a state (except for courts to enforce contracts that anybody can freely enter). Such beliefs indeed encourage entrepreneurs to take risks. They are half-truths, and also half-lies, but that drawback does not make them useless.[8] Business ideologists nonetheless neglect that about half of Hong Kong's people live in state-financed housing and receive other subsidies. They also downplay extensive evidence that cartels link particular grocery store chains, pharmacies, shippers, container-port operators, builders, land companies, and other firms in oligopolies that limit price competition. These consortia are not wholly coherent monopolies controlling the whole market, and there has recently been increasingly open evidence of tension between them; but they extract surplus rents that go into the territory's most important political networks. These have always been in businesses that can remain separate from the public responsibilities of the government.

The oligopolistic pattern is obvious to many Hong Kong residents. Perhaps it is unfair to focus on a single tycoon when several are involved, but Li Ka-shing is famous for being the largest owner of the Park'nShop grocery chain, and he also is reported to be a major shareholder in the only widespread alternative, the Wellcome chain. Few buyers in Hong Kong see these seriously competing, at least for retail sales to the wealthier half of the population. Smaller family-run grocery stores exist, and they tend to sell more products from China rather than imports from overseas. Two grocery chain startups by outsiders (a PRC consortium and Carrefours) have failed in recent years, apparently because the oligopoly was able to run them out of town. Li is also reported as a major owner of Watson's, the main drugstore chain. Some residents opine that this company forms a pharmaceuticals-and-cosmetics oligopoly with the other chain, Manning's. As with grocery retailing, the existence of smaller drugstores run by Chinese families provides no organized competition, since the largest retailers naturally have leverage over wholesalers.

The Li Ka-shing empire also extends to shipping and to Hong Kong's large container port through control of Hutchison Whampoa, to Hongkong Electric and other utilities, and to the land-and-construction business of the Cheung Kong corporation. Li's son Richard started Tom.com, an Internet service provider. Even before many dot.com companies ran into financial troubles, Richard Li diversified into other electronics and media sectors, taking over and renaming the phone company, and building a "Cyberport" of rentable space on one of Hong Kong island's few large patches of available reclaimed land. He also bought interests in newspapers and a 30 percent share of ATV, one of the main television channels. The Li empire does not seem to require a major bank (unlike *zaibatsu* in Japan or *chaebol* in Korea) because its parts loan to each other or go to the stock market for capital.

There may nonetheless be some oligopoly in Hong Kong's banking sector, too. Several PRC-associated banks naturally aid each other. Many local firms, such as the Bank of East Asia, have independent roots in traditional lineages, which does not preclude their coordination with other companies to prevent new competition or to raise the fees they can extract from the public. Such rents amount, in any country, to a kind of legalized "taxation" by an unofficial government composed of large businesses, and in Hong Kong, the government has always been very close to the tycoons.

The recognized state, since British times, has played a regulatory role among these crypto-political business networks. These have often conflicted with each other severely but quietly. After the Handover, criticism over utility rates was nonetheless publicized by one set of tycoons (including those running Hopewell, Wharf, and Sun Hung Kai) against Hongkong Electric (in which Li Ka-shing has major interests). The proposal to build a bridge connecting Hong Kong, Shenzhen, Zhuhai, and Macau has pitted a group of seven developers (including Hopewell, Sun Hung Kai, Wharf, and Great Eagle) against Hutchison Whampoa (Li). Then the Great Eagle chairman dropped out of the alliance and made an unwitting joke by saying that all these arguments about money had become "too political."[9]

Reality contrasts with the fantasy that Hong Kong is a purely free market. This type of planning reduces social policy funding, which goes to capital accumulation instead. Of course, such capitalization *is* social policy of a corporate, nonstate kind. It serves development functions, and it is administered by a combination of government bureaucrats and local tycoons. This kind of policy can justify technical education, although it militates against serving anybody who is deemed nonproductive even if needy. Hong Kong planning has been prodigious. Much of it has been coordinated by power networks outside the state.[10] It has seemed apolitical only because it has been mostly informal, unofficial, and indirect.

A further kind of planning began in 1983 when Britain announced the start of negotiations with China about Hong Kong's future. This temporarily destabilized the economy. So the Hong Kong dollar was first pegged to the U.S. greenback on 17 October of that year. The peg was one of the factors that encouraged a more modern structure of work in Hong Kong, spurring a shift from industrial to commercial operations within the territory as capital flowed into China. The main rationale for the peg, in public, is that it allows international traders to predict their costs and profits with reduced risk on short- and medium-term shipping contracts. This justification seems apolitical, merely instrumental. The long-term effect of the peg, however, has been to compel an upgrading of Hong Kong's labor force. By making Hong Kong unskilled workers' salaries higher than they would otherwise be in international terms, the peg gave local companies strong incentives to move their low-productivity jobs abroad, especially into China where Hong Kong capital became very welcome. The peg somewhat counterveiled the effects of indirect official subsidies to wages. Along with the Communist Party's tacit ending of China's inward-looking revolution, the peg outdated the Hong Kong government's stress on support for housing, water, and food. Liberal ideologists neglect that the HK dollar's peg (within a narrow band) to the greenback was set by no market, and it inflated the international value of the local currency. The peg raised unemployment rates among low-productivity HK laborers—a politically sensitive change. It also helped to create more high-productivity finance and marketing jobs in Hong Kong.[11]

A quickening of change in the types of labor that the HK economy required began in the early 1980s, when local firms safely began to pour great amounts of money into Guangdong. Blue-collar workers in Hong Kong became unemployed, as low-skill jobs "whooshed" across the border to Shenzhen, Dongguan, and much farther inland.[12] The city's economy was upgraded; it became more commercial and less industrial. This was Hong Kong's new plan, aimed at technical and commercial modernization, replacing the older pre-1980s plan whose goal was support of export industries. By 2000, at least half the employees of HK companies were in Guangdong. Almost seven-eighths (86 percent) of Hong Kong's own employees were by then in the tertiary service sector. The portion in Shanghai was still just 45 percent. (These portions had risen from half of Hong Kong's workers in 1980, but from just one-quarter of Shanghai's in that year.) Hong Kong's local deindustrialization, as the labor chapter noted earlier, "has yielded a labor surplus of almost half a million factory workers."

Shanghai also shifted somewhat from aid for industrialization to aid for commercialization. Shanghai differed from Hong Kong, however, because of its less hermetic border. Heavy industries had long found enough land to build near Shanghai (e.g., steel at Baoshan, chemical plants along the Huangpu, electrical machinery in Minhang), whereas Hong Kong lacked enough space for heavy industries. Shanghai's light-industrial factories received far more competition from new plants elsewhere in China than did its heavy industries. The heavy plants employ fewer laborers. Shanghai's commercialization may seem more dramatic than Hong Kong's because it began from a lower base before reforms—and Shanghai's deindustrialization, to the extent it occurred, has been in textiles and other light industries that hired many workers. In Hong Kong, public employment was all with the government, not with businesses. As late as 2000, 12 percent of all Shanghai regular employees worked for collectives, and another three-fifths (59 percent) still worked for state enterprises.

Policy making is centralized in Shanghai, but implementation is relatively decentralized. This presents an interesting academic documentation problem, because Shanghai policies are often clear and consistent (sometimes more so than Hong

Kong policies), but what actually happens in the East China metropolis is more variant. For example, the housing chapter mentions "underground real estate developments" by unlicensed brokers for squatters on public land. These might be less remarkable in Lima or Lagos. Such disjuncture between policy and behavior is taken as a larger problem in China mainly because writers more often imagine the state is the only power. The ideal of a central decision to decentralize decision structures is now touted as modern by both cities' governments, despite the obvious tension in this goal.

Many actors are involved, and they all have policies. The migration chapter finds that "Ironically, despite the proud association of the Hong Kong Government with the economic principle of *laissez-faire*, it is the socialist administration in Shanghai which exhibits a greater flexibility in control, allowing market forces to regulate the mix and intensity of inflow of the migrants." There is more flexibility in the East China metropolis also because of remnants of populist socialism there. By the same token, there is more uneven implementation. That only means the implemented policies are local, not central.

Another example of this difference may be found in schools. Hong Kong education, although it is administered through a wide variety of kinds of institution at the primary and secondary levels, is mostly funded by the government. In Shanghai, extrabudgetary and private money now pay for more education than in Hong Kong. The content of education policy consequently varies, not just because Shanghai citizens are on average poorer.

Shanghai's formally centralized policy-making structure, which still gives non-business Communist Party interests some influence, is evident in many of the city's public decisions. The 1996 startup of the Shanghai Housing Provident Fund, based on Singaporean models and requiring employers' contributions to accounts for individual employees, followed a pattern mooted but not copied in Hong Kong because businesses there did not want to pay, and they control the government.

The new chief executive in Hong Kong, even though he is a businessman too, sits in a chair that cannot make him entirely comfortable with this structure—several of his British predecessors also were not happy with it.[13] Chief Executive Tung Chee-hwa, in his bold speech on Handover Day in 1997 and also in his first policy address that October, proposed that tenants who rented public housing should be publicly financed to purchase the houses. He was inspired by Singapore, where the government builds housing and mandatory contributions to the Central Provident Fund pay for home ownership by families. In housing, health finance, and pension funding, too, the plenipotentiary chief executive has tried to bow to Hong Kong's low-tax traditions, which have created reticence about government coercion of business contributions even to individually owned employee accounts. When the Asian financial crisis struck, these plans were quietly shelved. Only much later did the government announce that the Home Ownership Scheme was "indefinitely suspended" for lack of money to pay for it.[14] The new governor had strong political reasons to propose this plan, however. His successors may still hope to get out of the housing business that the Hong Kong state inherited from British times. If land prices fall further in the future and the government moves to more dependence on the less wealthy half of Hong Kong's population, needing support from a larger group than the tycoons, this plan could reappear.

When official actions have boded to solve the problems of big businesses, however, they have been quickly taken. The Hang Seng Index fell in 1998; so this nominally anti-interventionist, ideologically capitalist government threw HK$120 billion into propping up the prices of local blue-chip stocks. This was no free market. It was socialism for the owners of wealth. The mixture in Hong Kong's system had never been clearer. The government did not, of course, incur responsibility for continuing management of the companies in which it invested. At least it did not do so as a public organization, but the largest local capitalists have always been Hong Kong's most influential citizens. Recently, a few of them have had to come out of economic hiding, because they have been needed to defend their policies in public. They have assumed more overt roles in the official wing of Hong Kong's administration, taking quasi-ministerial posts and advertising the state's business-oriented policies in media and in the Legislative Council. The "Principal Officials Accountability System" and increasingly open disputes between important sets of tycoons (over economic issues involving government regulation, such as electricity prices and plans to build a bridge to Zhuhai and Macau) suggested new developments in this city's politics.

Hong Kong's public spending at the turn of the millennium stood at 23 percent of GDP. This portion is low for a place with a first-world economy, but it is almost one-quarter of total local demand and is far from negligible. Land sales provided 54 percent of all Hong Kong government money as late as 1997–1998. That portion plunged to 26 percent just four years later (2001–2002). The government in fiscal years after 2000 ran large deficits. Caught between rising costs because of the economic recession, on one hand, and an inability to sell new land lest the price drop further, on the other, the government had to dip into reserve funds. Fortunately, these were large (over HK$400 billion) because of budget surpluses in many previous years.[15] As a long-term policy, however, this approach was not sustainable. Someday taxes on companies would surely rise.

Ideologies vs. Realities in Accumulating Capital and Financing Welfare

Colonial legacies in Hong Kong, like the quasi-colonial heritage in Shanghai, stressed the importance of financial rather than more broadly social capital.[16] Leaders such as Thatcher and Deng, or Tung Chee-hwa and Jiang Zemin, made business as such the main business of government. Welfare, health, elder care, education, and other social functions have been privatized whenever possible, especially in

Hong Kong. Social policies for collective support and long-term success have not been easy to "reinvent" as profit centers, even when government officials have been mainly interested in economic competitiveness.

The economic policy orientation of governors in both Shanghai and Hong Kong has given welfare as such a relatively low priority. Collective goods have been bought only if they boded to bring fairly immediate benefits to companies. "Active" policies to meet structural opportunities for economic upgrading have included labor retraining, the creation of new jobs, and help for people seeking work. "Passive" policies, such as unemployment compensation or help for the unemployable, have been sparsely funded in Shanghai and practically omitted as a government responsibility in Hong Kong.

Sometimes unemployed people can create their own jobs. Tea-egg sellers in Shanghai have done this. So have the food deliverers of the "Mama Zhuang Clean Vegetable Service," described in an earlier chapter. Some poverty, however, is stubbornly structural. In Hong Kong during 1998, a financial crisis year, the GDP fell by 8 percent. Unemployment rose to the lately unprecedented level of 8 percent, with an additional 3 percent "underemployed" by 2000. The causes did not lie in an onslaught of laziness among Hong Kong workers. They lay, instead, in the financial crisis and an outflow of low-skill jobs to Guangdong. The welfare chapter reports an interview with an unemployed Hong Kong man who said he was "over the age of forty, with elementary education, and with unwanted skills in unwanted industries."

Hong Kong's Labor Department runs a Local Employment Service that lists job vacancies and tries to match them with applicants. Private licensed agencies supplement these efforts. In Shanghai, which has a more open border, the state itself monopolizes headhunting and job-finding functions, partly as a residence control. The number of underemployed SOE workers who are "down from their posts" (*xiagang*) has nonetheless soared.

Shanghai and Hong Kong both offer evidence that the general health or recession of an urban economy determines the speed at which privatization reforms can be financed. Shanghai more than tripled the amount of residential space for its companies' employees in the 1990s. New space went especially to workers in collective and private firms. Those in state-owned enterprises were often allowed to keep their company flats at rents that remained concessionary. Most SOEs now lack money to loan so that employees might buy their homes and thus privatize housing.

In Hong Kong, the rate of housing privatization also varies by periods of boom or recession (although not by any difference in the ownership sector of the employees' jobs, which are overwhelmingly private). Property prices rose sharply from the end of 1995 to almost the end of 1997. Then they fell just as sharply—at least 60 percent on average by 2002, with predictions of further drops later. At the turn of the millennium, about 170,000 Hong Kong apartment owners had negative net worths.

One way to lower the portion of public housing in a city is to build less of it. The Hong Kong government limited new construction of public housing and sold some flats by the late 1980s. It tried to reallocate its available public housing space to needy tenants by doubling rents after 1988, or else by evicting, in 1999, households whose incomes were twice the Waiting List Income Limit.

In the East China metropolis, still nominally under a dictatorship of the proletariat, rents were also raised—but far more slowly than reformers wished, because workers in these homes remained a core constituency of the regime. Construction slowdowns in Shanghai near the turn of the millennium followed a building bubble of the early 1990s. These delays were unintended (and they reduced the urban immigration of construction workers), but Shanghai had much unrented new space. Antiplan reforms are themselves plans. Like other designs, they sometimes go awry.

In both cities, the municipal governments urgently hope to lessen the rates at which they manage housing. This is easier proposed than done. Hong Kong's government had built 52 percent of all the territory's dwellings by the turn of the millennium. It was still the direct landlord of 38 percent of all households. Despite the decrease of land prices and an ardent official will to get out of the rental business, the government has been able to sell an accumulated total of only 14 percent of all local homes to the residents. Executive Tung suggested he wanted to reduce his organization's exposure to this trade. Many public flats were deteriorating, however, and the rent rates were still low enough to discourage renters from buying their flats, despite the falling values of the apartments.

Managing medicine and elder care were officially just as suspect as managing residential estates. Hong Kong has a "national" health care system, at least for secondary and hospital care. It was set up gradually, in decades when the city's population was young on average and less prone to need money to cure expensive maladies. The latest medical technology at that time was less costly than it has recently become. Demand curves for some kinds of medicine are essentially vertical; people will buy, if they can, irrespective of price. If they cannot, they want the state to help pay. Government provision of health care retained the minimalist style of the colonial business-run regime. Medium and small companies, of which Hong Kong still has many, could not afford specialized medical departments for their employees. A compromise was reached under which individuals paid for primary care and the state spread, to Hong Kong's reluctant taxpayers, the financial risks of major medical care that few people need until old age. In Shanghai, however, where small and large firms alike were often grouped under the oversight of municipal bureaus, primary care of quite basic quality has been generally funded for employees and their families. In both Shanghai and Hong Kong now, there is little political will either to forgo costly new medical technologies or to pay for them.

A similar governmental unwillingness to spend applies to elder care. Hong Kong's institutions for the aged are themselves old. The most important is entirely traditional: the Chinese family. The territory has far more money than Shanghai does per senior citizen, but many frail old people in Hong Kong nonetheless do not survive the waiting time and income qualifications to get into the system that is partly financed by the local state. The average delay, for applicants in 2001 to Hong Kong's "care and attention homes," was nearly three years (thirty-four months). The government gave nongovernmental organizations some money to provide limited care and nursing in these homes, although it understandably preferred families to take on the burden of elder care. As the relevant chapter notes, the waiting list for public help includes about 30,000 elderly people: "Therefore, quite a number of the applicants die while still awaiting admission." A major message here, for both Hong Kong and Shanghai, is demographic. The portion of old people in both cities is quickly accelerating. Their sheer number threatens to swamp the limited finances budgeted to nurse them.

Less expensive realities, however, still capture business-governmentalists' attention. Depending on cost, compromises can be reached between the ideologists of capital growth and welfare protection. Because Hong Kong's health service is government-run, and because cigarettes (not manufactured in the territory) cause massive long-term costs, the antismoking campaign gets HK$30 million annually. Shanghai apparently did not match this wise investment.

Health finance reforms introduced in Shanghai took a Singaporean form after these policies were confirmed by mid-1990s experiments in Zhenjiang and Jiujiang. Mandatory contributions, mainly from employers, go into employees' individual accounts for secondary and tertiary care. Except for catastrophic illnesses, additional expenses are mostly paid out of pocket. A prospering economy is required to make this method of finance adequate and to assure that few are unemployed and thus left with no insurance. This system does not, however, automatically include Shanghai's workers in private or collective firms. Primary care comes mostly from general practitioners and private clinics, paid directly by patients.

Greater employer and employee contributions to Hong Kong's health care finances were proposed in 1999, and politicians of practically all parties balked at these new quasi-taxes. As Anthony Cheung, the former Legco member and leader of the Democratic Party who has written the health chapter notes, "the politics of economic recession" has delayed sustainable financing that might carry the current norms of universal availability of medical care into the future. When governments are run as businesses, their executives naturally lay off the needy, the old, and the sick. Money for welfare does, indeed, decrease the short-term rate of capital accumulation.

Metropolitan Adjustments to Inland Development

Both Shanghai and Hong Kong are losing out if they are conceived as separate from China's many booming middle-sized cities. Some residents in middling cities nonetheless regard these metropolises as modern rather than alien. Big-city administrators, worried about competitiveness, often have a different viewpoint from that of many ordinary people. Each of the two cities examined here is deeply affected by its Chinese context.

Hong Kong and Shanghai people spend much time discussing the competitive threat posed by the other metropolis. Yet these are two obviously similar places. The Cantonese city worries that Shanghai will become the main financial center in the Far East—as indeed it might. This reverses Hong Kong's previous fortune in having avoided the violence of Mao's revolution, because it was a British colony. Historically, many Cantonese played major roles in Shanghai's early development. Shanghai emigrant entrepreneurs were likewise pivotal in Hong Kong's industrialization.[17] These places are more like each other than either is like Beijing. They compete with each other, and also with midsized cities and with many ports along the coast from Zhanjiang through Xiamen to Qingdao. Their economic watersheds nonetheless remain large enough that they can both prosper, despite this competition and the adjustments they undergo.

Hong Kong shifted to a "knowledge-based economy" by the 1990s, as described earlier. Shanghai's recent transition has been more diverse, but such changes in any case cannot be assessed just by looking at the municipalities alone. Hong Kong's change is linked to the rise of light industries that the city's capitalists financed throughout the Pearl River Delta and elsewhere. Without the marketing and accounting expertise of Hong Kong offices, factories on the delta could not have sold so internationally or prospered so greatly.

To see Shanghai's change, it is also necessary to look at the whole Yangzi Delta, where high-technology industries have burgeoned in cities such as Wuxi, Suzhou, Jiaxing, Kunshan (which is in Jiangsu but barely across Shanghai's border), and Ningbo (whence at least one-quarter of Shanghai's managerial families originally came). Each of these cities has its own economic elite—and each is connected to Shanghai by many informal ties and formal joint-venture arrangements. New electronics technologies have been applied successfully in Wuxi factories, for example, by entrepreneurs who are also Shanghai people and often use Shanghai capital. New transport technologies have similarly been adopted in Ningbo. These Jiangnan cities developed in ways that are inseparable from changes in the metropolis. None of them are province-level capitals, and all enjoy lower taxes and less strenuous monitoring than Shanghai, Nanjing, or Hangzhou. They compete with Shanghai in some ways, are coordinated with Shanghai in others, and are largely run by people who identify themselves as Shanghainese, too.

Many of "Hong Kong's" elite likewise have allegiances (and financial interests) in several Chinese cities. The most famous of all the current tycoons, Li Ka-shing, periodically threatens to take his money back to Ningbo, whence he came, whenever he smells too much democracy in Hong Kong. This could, after all, raise his taxes.

City-limit sociology or economics cannot capture what has been happening in either the Pearl River or Yangzi River Deltas. Insufficient attention to the contexts of cities has caused many inadequate interpretations of urban change. The problems that Hong Kong has faced since 1997, for example, are not the difficulties that many pessimists predicted. Dire auguries were pronounced before the Handover, but they seldom included guesses that the land price would collapse. No change has been more important to more families than this. Apparently no pre-1997 guru predicted flats would lose half their value. This deflation hurt the soundness of banks and of the government budget. Some prognosticators foresaw popular anger, but not that it would arise because so many would lose half their wealth. Nobody said the government would be worried about its ratings in polls for that kind of reason. Problems were expected to be political, not economic.

Even as regards politics, those who were killed in the 1989 Tiananmen catastrophe are now still remembered in a candlelight vigil that is held each year at Victoria Park. Freedom of public speech in Hong Kong has thus far been maintained, despite conflicting gestures of patriotism and of liberalism that surround the antitreason clause, Article 23, of the Basic Law. Timidity among journalists has come less from government action than from media purchases, which have clarified that free markets do not always aid other freedoms. Nathan Road has not yet been rechristened Liberation Road (the name of the main entryway to most other large Chinese cities). The SAR's rights of final adjudication have been abridged—to exclude poor outsiders, a cause that populist sometime-democrats supported shamelessly. Nobody guessed this government would be unable to handle smoothly a quasi-administrative, technical-scientific problem like the bird flu.

The successes of the SAR, likewise, have not been what the optimists predicted before 1997. Many patriots thought that the one sure bet for this city within China would be increased economic prosperity. Hong Kong is adjusting to the news that it is definitely not the only usable port on the China coast. Profits from Hong Kong capital have been reduced by competition from other PRC harbors. Economic growth in the territory has fallen. In the second quarter of 2000, nearly half a million Hong Kong workers earned less than HK$6,000 (U.S.$770) per month— and this was a 15 percent rise, in just one year, of the size of the lower class who were impoverished at that level.[18] It is natural that Hong Kong is in China, but that has not solved all problems. Fewer than five years have disproved most pre-Handover hopes and warnings alike.

The Beijing context also affects both Hong Kong and Shanghai. Many of China's top recent leaders know Shanghai well; they spent parts of their careers there. For Hong Kong, too, Beijing is crucial, although the national regime may someday represent Cantonese more fully. Constitutional constraints on Hong Kong's oddly constructed Legislative Council have discouraged many local democrats, but a weakened party system still exists. Practically all power is invested in the chief executive, but even he now relies on an Executive Council, including heads of the "patriotic" parties of business (the Liberals), lower-income residents (the Democratic Alliance for the Betterment of Hong Kong), and the Beijing-approved Federation of Trade Unions.

The social policies that benefit groups voting for these "patriotic" parties, on which the government depends, are far from identical.[19] It is easy to name politicians whom Beijing likes and politicians whom Beijing dislikes among representatives of both capital and labor, both right and left. So what policies are actually decided? On any question—land price, elder care, health funding, education, unemployment, or any other issue requiring an official decision—government parties are no more alike than are the constituents of democrats, who also range from housing estates to executive suites. Leadership by a chief executive makes the job of finding good policy no more straightforward than would any other kind of leadership. The Basic Law in any case provides that the nearly omnipotent chief executive should eventually be elected by universal suffrage. Without a major change in Beijing, this is unlikely to happen by the end of the next term in 2007. If it does not, the issue of popular participation will surely recur in 2012, 2017, or later, until it is settled. When that happens, the political reasons for Legco's constitutional weakness will disappear, and the political parties may evolve to make decisions, not just debates.

The main use of politics is to legitimate official actions by giving their implementers a sense of participation in them, or at least a sense that the actions are rightfully taken. The idea that governments should always follow markets inherently contradicts this function of politics, because purely economic exchanges produce unintended results; yet politics are all about intended understandings. Markets coordinate resources and people efficiently, but in a mechanical rather than a humanly monitored way. To the extent that they are regulated, they are not free markets. The most obvious result of a market is the finding of a price—which is unintended by any relevant actor (the seller wants it higher; the buyer wants it lower).

Several chapters contrast "efficiency" as determined by markets with its version as determined by managers who claim to know what markets would prescribe. Any managerial interpretation of "accountability" is unadmitted planning. Socialists can pose as liberals, and vice-versa, because the difference between them is less than the difference between developed rich situations and undeveloped poor ones. The former, which are partly based on efficient market coordinations of resources, can alone provide the tax base both to enforce free contracts and to ensure social welfare. Poor places, which lack money, have for that reason less success in achieving the goals of either individuals or collectives. The education chapter contrasts "Hong Kong . . . corporate management" with "Shanghai . . . market forces and new initiatives from the non-state sectors." This is One Country, Two Systems—but reversed. Much evidence throughout this book disproves claims that Hong Kong is just capitalist or Shanghai just planned.

Modernity as Mixed: Capitalization with Compassion

A current goal in both Hong Kong and Shanghai is to transform centralized state control into decentralized market decisions. Actual behavior in both of these cities makes clear, however, that there is a great deal of government in business and a great deal of business in the state. Whether these should ideally intertwine so much, in fact they do. As the chapter on education suggests, "one should not analyze these processes of 'marketization' and 'decentralization' as simply a one-dimensional shift from the 'state' (as bureaucratic and nonmarket) to the 'market' (as corporate and nonstate)." Not only do midsize social units make at least as many important decisions as whole polities or rational individuals, they also implement these decisions. Fashionable jargons about "competitiveness," "accountability," or the "three E's" (economy, efficiency, and effectiveness) are mobilization slogans, not sufficient policies.

Ideological discourses about capital accumulation or social compassion are not solely about collective problems. They are also existential, about images of what a modern person should be. The official image in Hong Kong and Shanghai is that a good person is productive and active, not resigned. It is safest, in this view, to be specialized, even one-track or obsessive, to guarantee not being lazy. Some personalities are more competitive than others, and some are admirably so. Existential choices of personal ideals deeply affect notions of proper social policy. In practice, people differ. The opportunity for good policy is to see that real problems are also varied.[20] The aim is to solve them concretely, rather than to pretend that mere discovery of an idea (either individualist or communitarian) would be enough to spur enterprise and to make people happy.

Notes

1. For evidence about this, see L. White, *Unstately Power: Local Causes of China's Economic Reforms* (Armonk, NY: M.E. Sharpe, 1998).
2. See http://cf.heritage.org/index/indexoffreedom.cfm (November 11, 2002).
3. For examples, see Elizabeth Sinn, *Power and Charity* (Hong Kong: Oxford University Press, 1989) about the Tung Hwa Hospital, or Barbara-Sue White, *Turbans and Traders* (Hong Kong: Oxford University Press, 1994) about the initiative of a Parsee businessman in the founding of Hong Kong's first university.
4. A new sociological classic about the Western comparisons is Frank Dobbin, *Forging Industrial Policy: The United States, Britain, and France in the Railway Age* (Cambridge: Cambridge University Press, 1994).
5. The factors that created this path-dependency (history) in Japan since the 1920s are famously described in Chalmers A. Johnson, *MITI and the Japanese Miracle: The Growth of Industrial Policy, 1925–1975* (Stanford, CA: Stanford University Press, 1982).
6. See Chapter 3 on housing. This high figure comes from 1997–1998, but the portion in previous years was also very large.
7. See Manuel Castells, Lee Goh, and Reginald Y.W. Kwok, *The Shek Kip Mei Syndrome: Public Housing and Economic Development* (London: Pion, 1991), which refers to Jonathan Schiffer, "The State and Economic Development: A Note on the Hong Kong 'Model,'" *International Journal of Urban and Regional Research* 15 no. 3 (1991). I am indebted to Professor Jeffrey Henderson for these references. The exact econometrics behind such estimates is subject to discussion, but the main point (that Hong Kong's economy was massively planned in the industrializing period, because of indirect government subsidies to firms' wage bills) has been thoroughly documented and widely ignored.

8. The classic on the functions of social consciousness is Karl Mannheim, *Ideology and Utopia: An Introduction to the Sociology of Knowledge* (London: Routledge, 1954).

9. The author has benefited from diverse conversations with many, including John Walden (who describes sharp but unpublic tensions under the British between the Hongkong and Hang Seng banks, before the former bought out the latter), Nailene Chou Weist, and other helpful Hong Kong observers. On the bridge and electricity issues, see Christine Loh, "Politics in High Gear," www.clsa.com, p. 13.

10. On distortions that arise from a willingness to find politics only in government institutions rather than businesses, see E.E. Schattschneider, *The Semi-Sovereign People* (Hinsdale, IL: Dreyden Press, 1975). An index of change is that Li Ka-shing now denies in public that he is anything like a monopolist; see http://latelinenews.com/ll/english/1073062.shtml (November 12, 2002).

11. For some of these ideas the author is indebted to Lee Pui-tak of the Centre of Asian Studies, Hong Kong University, and also to a presentation by Christine Loh at the Association for Asian Studies Annual Meeting in Washington, DC, on 5 April 2002. The "peg" is really a band around the approximate value U.S.$1 = HK$7.8. The government directly intervenes on this market, to buy or sell, when the price veers below or above a narrow range.

12. The populist U.S. third-party presidential candidate Ross Perot, who opposed the North American Free Trade Agreement, described "that whooshing sound" of American jobs flowing into Mexico, especially to *maquiladora* factories just south of the Texas and California borders. In proportional terms, the drain of blue-collar jobs from Hong Kong was far more severe, but few politicians were far enough from business interests even to suggest that the government might do anything about it.

13. Patten, Maclehose, and to some extent others as far back as governors with experience in Ireland (Blake and Pope-Hennessey) had social and political policies that some businesses disliked. See Lynn White, "The Appeals of Conservatives and Reformers in Hong Kong," in *Hong Kong Reintegrating with China*, ed. Lee Pui-tak (Hong Kong: University of Hong Kong Press, 2001), pp. 3–38.

14. See Christine Loh, "Politics in High Gear," at www.clsa.com, p. 5.

15. See Joseph Yu-shek Cheng, "The Challenges of the HKSAR Government in Times of Economic and Political Difficulties," *American Asian Review* 20, no. 1 (spring 2002): 42.

16. On Shanghai, see L. White, "Non-Governmentalism in the Historical Development of Modern Shanghai," in *Urban Development in Modern China*, ed. Laurence J.C. Ma and Edward W. Hanten (Boulder, CO: Westview Press, 1981), pp. 19–57.

17. Wong Siu-lun, *Emigrant Entrepreneurs: Shanghai Industrialists in Hong Kong* (Hong Kong: Oxford University Press, 1988).

18. Hong Kong's 205,000 foreign domestic servants were not included in the 450,000 relatively poor people described earlier. See Cheng, "The Challenges," p. 37.

19. Ibid., p. 53, notes proposals that DAB chair Tsang Yok-sing made in 2000, favoring a ministerial system and a greater official role for pro-Beijing parties. Tsang has an unenviable job. He needs to support government policies that favor capital while garnering votes for the DAB from labor.

20. For example, despite the aging of Hong Kong, the territory's overall dependency ratio is nearly the world's lowest. Financing dependents need not have been seen by Hong Kong officials as a big problem. There is money to pay for them. The unwillingness to spend it freely may show more about the increasing competitiveness of inland places and about ideals of personality than about current social problems. See United Nations Development Program, *Human Development Report 2000* (New York: Oxford University Press, 2000), p. 223, which compares dependency in Hong Kong and in 174 UN members for 1998.

Bibliography

Agelasto, M., and B. Adamson. *Higher Education in Post-Mao China*. Hong Kong: Hong Kong University Press, 1998.

Aleinikoff, T.A., and D. Klusmeyer, eds. *From Migrants to Citizens: Membership in a Changing World*. Washington, DC: Carnegie Endowment for International Peace, 2000.

Arnsberger, P., P. Fox, X.L. Zhang, and S.X. Gui. "Population Aging and the Need for Long Term Care: A Comparison of the United States and the People's Republic of China." *Journal of Cross-Cultural Gerontology* 15 (2000): 207–27.

Aucoin, P. "Administrative Reform in Public Management: Paradigms, Principles, Paradoxes and Pendulums." *Governance* 3 (1990): 115–37.

Ball, S. "Big Policies/Small World: An Introduction to International Perspectives in Education Policy." *Comparative Education* 34 (1998): 119–30.

Baltodano, A. "The Study of Public Administration in Times of Global Interpretation: A Historical Rationale for a Theoretical Model." *Journal of Public Administration Research and Theory* 7 (1997): 623–26.

The Basic Law of the Hong Kong Special Administrative Region of the People's Republic of China. The Heritage Foundation, *2001 Index of Economic Freedom*. Washington, DC: Heritage Foundation and *Wall Street Journal*, 2000.

Bates, R. "The Educational Costs of Managerialism." Paper presented to the Joint Conference of the Educational Research Association, Singapore and the Australian Association for Research in Education, Singapore Polytechnic, Singapore, 25–29 November 1996.

Beijing Review, Beijing, various issues.

Bellman, T. *Housing Reform: The Engine of Growth*, Powerpoint Slides, 14 May 2002. www.joneslanglasalle.com.hk/Research/2002/HSBC%20Shanghai%20Conf%20%20May2002.pdf.

Berger, S., and R.K. Lester, eds. *Made by Hong Kong*. Hong Kong: Oxford University Press, 1997.

Bergere, Marie-Claire. *The Golden Age of the Chinese Bourgeosie 1911–1937*. Cambridge: Cambridge University Press, 1989.

Betcherman, G., A. Dar, A. Luinstra, and M. Ogawa. *Active Labor Market Programs: Policy Issues for East Asia*. Washington, DC: World Bank Group, 2000.

Blundell, C. "Housing: Getting the Priorities Right." In *Managing the New Hong Kong Economy,* edited by D. Mole, 108–49. Hong Kong: Oxford University Press, 1996.

Bo, X.F., and J.Z. Dong. "*Yiliao weisheng gaige jubu weinan: guanyu woguo yiliao weisheng gaige de diaocha*" (Medical and Health Reforms' Difficult Maneuvers: A Survey on Our Country's Medical and Health Reforms) (Part 1). In *Zhongguo Jingji Tizhi Gaige* (China's Economic Structure Reform) 1 (1993): 56–58.

Bonavia, D. *Hong Kong 1997: The Final Settlement*. Hong Kong: SCMP, 1985.

Bray, M. *Privatisation of Secondary Education: Issues and Policy Implications*. Paris: Division of Secondary Education, UNESCO, 1996a.

Bray, M. *Counting the Full Cost: Parental and Community Financing of Education in East Asia*. Washington, DC: World Bank, 1996b.

Bray, M. "Financing Higher Education in Asia and the Pacific: Patterns, Trends and Options." Paper presented at the World Congress on Higher Education and Human Resource Development in Asia and the Pacific for the 21st Century, Manila, 23–25 June 1997.

Bray, M. "Control of Education: Issues and Tensions in Centralization and Decentralization." In *Comparative Education: The Dialectic of the Global and the Local*, edited by R.F. Arnove and C.A. Torres, 207–32. Lanham, MD: Rowman and Littlefield, 1999.

Bridges, D., and J. McLaughlin, eds. *Education and the Market Place*. London: Falmer Press, 1994.

Brown, D.J. *Decentralization and School-Based Management*. London: Falmer Press, 1990.

Cai, Laixing, ed. *Shanghai: Building a New International Economic Center*. Shanghai: Shanghai People's Press, 1995.

Calavita, K. "U.S. Immigration and Policy Responses: The Limits of Legislation." In *Controlling Immigration: A Global Perspective*, edited by W. Cornelius, P. Martin, and J. Holliefield, 56. Stanford, CA: Stanford University Press, 1994.

Castells, M., L. Goh, and R.Y.W. Kwok. *The Shek Kip Mei Syndrome: Public Housing and Economic Development*. London: Pion, 1991.

Census and Statistics Department, HKSAR Government. *Social Data Collected by the General Household Survey: Special Topics Report No. 15*. Hong Kong: Printing Department, Hong Kong Government, 1997.

Census and Statistics Department, HKSAR Government. *Hong Kong Life Tables (1996–2031)*. Hong Kong: Government Press, 2000.

Census and Statistics Department, HKSAR Government. *Social Data Collected via the General Household Survey: Special Topics Report No. 27*. Hong Kong: Government Press, 2001.

Census and Statistics Department, HKSAR Government. Census and Statistics Department Homepage at www.info.gov.hk/censtatd/eng/hkstat/fas/01c. Accessed on 26 July 2002.

Central Intelligence Agency. *The World Fact Book 2000*. Washington, DC: Central Intelligence Agency, 2001.

Cerny, P. "Paradoxes of the Competition State: The Dynamic of Political Globalization." *Governance and Opposition* 3 (1996): 251–74.

Chan, D. "Marketization and Educational Developments in China." *World Studies in Education* (2001): 2.

Chan, D., and K.H. Mok. "Educational Reforms and Coping Strategies under the Tidal Wave of Marketization: A Comparative Study of Hong Kong and the Mainland." *Comparative Education* 37 (2001a): 21–41.

Chan, D., and K.H. Mok. "The Resurgence of Private Education in Post-Mao China: Problems and Prospects." In *Education, Culture and Identity in Twentieth Century China*, edited by G. Peterson, R. Hayhoe, and Y.L. Lu. Ann Arbor: University of Michigan Press, 2001b.

Chan, D., and K.L. Ngok. "Centralization and Decentralization in Educational Development in China: The Case of Shanghai." *Education and Society* 19, no. 3 (2001): 59–78.

Chan, E. "Defining Fellow Compatriots as 'Others'—National Identity in Hong Kong." *Government and Opposition* 35, no. 4 (autumn 2000): 499–519.

Chan, R.K.H. *Welfare in Newly Industrialised Society: The Construction of the Welfare State in Hong Kong*. Aldershot: Avebury, 1996.

Chao, Y.Y. *The 21st Century Private Higher Education in the Mainland.* Taipei: Taiwan University Press, 2001.

Chen, F. "The Re-employment Project in Shanghai: Institutional Workings and Consequences for Workers," *China Information* 14, no. 2 (2000).

Chen, M.Z. *"Zhongguo wiesheng shiye gaige yu fanzhan de huigu yu zhanwang"* (China's Health Care Reform and Development: Retrospect and Prospect). Special Address at the Hospital Authority Convention on "Reinventing Health Care for the 21st Century," Hong Kong Hospital Authority, Hong Kong, 11–13 May 1997.

Chen, Y., ed. *Dangdai Zhongguo de Shanghai* (Shanghai in Contemporary China), Vols. I and II. Beijing: Dangdai Zhongguo Chubanshe, 1993.

Cheng, J., ed. *A Chronology of the People's Republic of China 1949–1984.* Beijing: Foreign Languages Press, 1986.

Cheng, K.M. "Education-Decentralization and the Market." In *Social Change and Social Policy in Contemporary China,* edited by L. Wong and S. McPherson. Avebury: Aldershot, 1995.

Cheng, T., and M. Selden. "The Origins and Social Consequences of China's *hukou* System." *China Quarterly* 139 (1994): 644–68.

Cheng, X.M., and X.F. Ye. *"Shanghai xian yiliao weisheng shoufei"* (Shanghai County Medical and Health Charges). *Shanghai diyi yixueyuan xuebao zengkan* (*Shanghai First Medical School Journal,* supplement) (1982): 91–95.

Cheung, A.B.L. "Health Policy Reform." In *The Market in Chinese Social Policy,* edited by L. Wong and N. Flynn, 63–87. Basingstoke: Palgrave, 2001.

Cheung, P.T.Y. "The Political Context of Shanghai's Economic Development." In *Shanghai: Transformation and Modernization under China's Open Policy,* edited by Y.M. Yeung and Y.W. Sung. Hong Kong: Chinese University Press, 1996.

China National Committee on Aging Problems. *The Imminent Population Aging in China.* Beijing: China Labor Science Editorial Office, 1988 (in Chinese).

China Shanghai, Official Web site of Shanghai Municipality, www.shanghai.gov.cn/gb/shanghai/English/BasicFacts/ (2001).

China Statistical Bureau. *Zhongguo Tongji Nianjian 1998* (China Statistical Yearbook 1998). Beijing: China Statistical Publishing House, 1998.

China Statistical Bureau. *Zhongguo Tongji Nianjian 2000* (China Statistical Yearbook 2000). Beijing: China Statistical Publishing House, 2000.

China Statistical Yearbook 1998, 2001. Beijing: China Statistical Bureau, 1998, 2001.

Chinese Communist Party Central Committee (CCPCC). *The Decision of the Central Committee of the Communist Party of China on the Reform of Educational Structure.* Beijing: People's Press, 1985.

Chinese Communist Party Central Committee (CCPCC). *The Program for Reform and Development of China's Education.* Beijing: People's Press, 1993.

Chinese Labor Statistical Yearbook. Beijing: Chinese Statistical Bureau, 1997.

Chiu, R.L.H. "Housing." In *Shanghai: Transformation and Modernization under China's Open Policy,* edited by Y.M. Yeung and Y.W. Sung, 341–74. Hong Kong: The Chinese University Press, 1996.

Chong, A.M.L. "Residential Care for the Elderly in Urban Areas: A Future Challenge for China." In *Social Welfare Development in China,* edited by T.W. Lo and J.Y.S. Cheng, 179–96. Chicago: Imprint Publications, 1996.

Chong, M.L., and Y.H. Kwan. *A Study of the Satisfaction, Intergenerational Relationship, and Decision Making of the Institutionalized Elderly—A Comparison of Hong Kong, Guangzhou, Shanghai and Beijing.* Hong Kong: Department of Applied Social Studies, City University of Hong Kong, 2001.

Chow, N. *Socialist Welfare with Chinese Characteristics: The Reform of the Social Security System in China.* Hong Kong: Centre of Asian Studies, University of Hong Kong, 2000.

Chow, N.W.S. *The Development of Social Welfare Policy in Hong Kong.* Hong Kong: University Press Company, 1980.

Chow, N.W.S. *Xianggang Shehui Fuli Zhengce Pingxi* (Comment and Analysis of Hong Kong Social Welfare Policy). Hong Kong: Tiandi Tushu Youxian Gongsi, 1984.

Chu, P. *Programmes and Services of Long-Term Care in Hong Kong.* In *Conference Proceedings of the Workshop on Long Term Care for the Elderly at the Turn of the New Century: An International Experience.* Organized by the Centre of Aging of the University of Hong Kong and Cosponsored by the Elderly Commission, Hong Kong, 11–12 January 2000.

Clarke, J., A. Cochrane, and E. McLaughlin, eds. *Managing Social Policy.* London: Sage, 1994.

Commissioner for Labour. *1999 Report of the Commissioner for Labour.* Hong Kong: Government Printer, 2000.

Commission of the European Communities. *Social Protection in Europe 1995.* Brussels: European Commission, 1995.

Cong, C. "The Strategy of Developing the Housing Provident Fund System—The Case of Shanghai." *Chinese and Foreign Real Estate Times* 8 (2001): 12–13.

Consumer Council. *How Competitive is the Property Market?* Hong Kong: Consumer Council, 1996.

Cui, P. Outline of Presentation in the seminar on "The Development of Housing Policy, Real Estate Policy and Property Market in Shanghai," Department of Public and Social Administration, City University of Hong Kong, Hong Kong, 5 July 2002.

Curriculum Development Council (CDC). *A Holistic Review of the Hong Kong School Curriculum: Proposed Reforms.* Hong Kong: Hong Kong Government Printing Department, 1999.

Currie, J., and J. Newson, eds. *Universities and Globalization: Critical Perspectives.* Thousand Oaks, CA: Sage Publications, 1998.

Dale, R. "The State and the Governance of Education: An Analysis of the Restructuring of the State–Education Relationship." In *Education: Culture, Economy and Society,* edited by A.H. Halsey, H. Launder, P. Brown, and A.S. Wells. Oxford: Oxford University Press, 1997.

Dale, R. "Specifying Globalization Effects on National Policy: A Focus on the Mechanisms." *Journal of Education Policy* 14 (1999): 1–17.

Davis, D. "Job Mobility in Post-Mao Cities." *China Quarterly* 132 (1992): 1062–85.

Deem, R. "Globalisation, New Managerialism, Academic Capitalism and Entrepreneurialism in Universities: Is the Local Dimension Still Important?" *Comparative Education* 37 (2001): 7–20.

Department of Social Welfare, Shanghai Bureau of Civil Affairs. *Annual Report of Social Welfare in Shanghai—2001.* Shanghai: Shanghai Municipal Bureau of Civil Affairs, 2002 (in Chinese).

Dobbin, F. *Forging Industrial Policy: The United States, Britain and France in the Railway Age.* Cambridge: Cambridge University Press, 1994.

Dong, F. "Current Position and Problems of Housing Allocation System Reform in Some Shanghai State-Owned Enterprises and Institutions." *Housing and Real Estate* 53 (August 1999): 9–12.

Dudley, J. "Globalization and Higher Education Policy in Australia." Paper presented at the 9th World Congress of Comparative Education, The University of Sydney, Sydney, Australia, 1–6 July 1996.

Editorial Board, China Education Yearbook. *China Education Yearbook 1997.* Beijing: People's Education Press, 1997.

Education and Manpower Branch (EMB). *The School Management Initiative: Setting the Framework for Quality in Schools.* Hong Kong: Hong Kong Government Printing Department, 1991.

Education and Manpower Bureau (EMB). *Information Technology for Quality Education: 5-Year Strategy: 1998/99 to 2002/3.* Hong Kong: Hong Kong Government Printing Department, 1998a.

Education and Manpower Bureau (EMB). *Review of the Education Department Consultation Document.* Hong Kong: Hong Kong Government Printing Department, 1998b.

Education and Manpower Bureau (EMB). *Information Technology for Learning in a New Era: Five-Year Strategy 1998/99 to 2002/3.* Hong Kong: Hong Kong Government Printing Department, 1998c.

Education Commission (EC). *Education Commission Report No. 7: Quality School Education.* Hong Kong: Hong Kong Government Printing Department, 1997.

Education Commission (EC). *Establishment of a General Teaching Council Consultation Document.* Hong Kong: Hong Kong Government Printing Department, 1998.

Education Commission (EC). *Education Blueprint for the 21st Century, Review of Academic System: Aims of Education Consultation Document.* Hong Kong: Hong Kong Government Printing Department, 1999a.

Education Commission (EC). *Education Blueprint for the 21st Century, Review of Education System, Framework for Education Reform: Learning for Life.* Hong Kong: Hong Kong Government Printing Department, 1999b.

Education Commission (EC). *Education Blueprint for the 21st Century, Review of Education System Reform Proposals Consultation Document.* Hong Kong: Hong Kong Government Printing Department, 2000a.

Education Commission (EC). *Reform Proposals for the Education System in Hong Kong.* Hong Kong: Hong Kong Government Printing Department, 2000b.

Education Department (ED). *Quality Assurance in School Education: Performance Indicators (Primary).* Hong Kong: Hong Kong Government Printing Department, 1998a.

Education Department (ED). *Quality Assurance in School Education: Performance Indicators (Secondary).* Hong Kong: Hong Kong Government Printing Department, 1998b.

Education Department (ED). *Performance Indicators in the Domain of Teaching and Learning (Primary),* Revised edition. Hong Kong: Hong Kong Government Printing Department, 2000a.

Education Department (ED). *Performance Indicators in the Domain of Teaching and Learning (Secondary),* Revised edition. Hong Kong: Hong Kong Government Printing Department, 2000b.

Education Department (ED). *Transforming Schools into Dynamic and Accountable Professional Learning Communities: School-Based Management Consultation Document.* Hong Kong: Hong Kong Government Printing Department, 2000c.

Employees Retraining Board. *1998–1999 Annual Report.* Hong Kong: Employees Retraining Board, 2000.

Employees Retraining Board Web site, 2000 (www.erb.org/english/index8.html).

Encyclopaedia of Chinese Social Work. Beijing: Chinese Society Press, 1994 (in Chinese).

Endacott, G.B. *Government and People in Hong Kong.* Hong Kong: Hong Kong University Press, 1964.

Fangdichan Yanjiu yu Dongtai (Housing and Real Estate Studies and Dynamics), no. 2 (2002).

Faure, D., ed. *Society: A Documentary History of Hong Kong.* Hong Kong: Hong Kong University Press, 1997.

Feng, G.S. "Yingjie 21 Shiji Laolinghua de Tiaozhan." In *Maixiang 21 Shiji Laoling Wenti Guoji Yantaohui Lunwenji* (Proceedings of the International Conference on Population Aging Issues in the 21st Century), edited by Guishan Feng, 1–15. Shanghai: Shanghai Kexue Jixu Wenxian Chubanshe, 1998.

Flynn, N. *Public Sector Management,* 3d ed. London and New York: Prentice Hall/Harvester Wheatsheaf, 1997.

Forrest, R., and P. Williams. "Future Directions?" In *Directions in Housing Policy: Towards Sustainable Housing Policies for the UK,* edited by P. Williams, 202. London: Paul Chapman Publishing, 1997.

Foster, C.D. *Privatization, Public Ownership and Natural Monopoly.* Oxford: Blackwell, 1992.

Fudan University Research Team on Re-employment Service Centres. *Research Report on Re-employment Service Centers and Projects.* Shanghai: Fudan University Press, 1998.

Fung, B.Y. *A Century of Hong Kong Real Estate Development.* Hong Kong: Joint Publishing (HK) Ltd., 2001, p. 297.

Gao, Q. "Waishengshi Qianru Liuru Renkou Xianzhuang Fenxi ji Sikao" (Analysis and Reflection on Population Inflow and Immigration from other Provinces and Cities). *Renkou* 3 (1997): 46–53.

Gao, Y.F. "Gao Yufeng zhuren zai gongjijin zhidu jianli shizounian jinianhui de jianghua tiyao" (Synopsis of the speech by Director Gao Yufeng in the tenth anniversary of the housing provident fund). *Zhongguo Zhufang yu Fangdichan Xinxiwang,* 25 April 2001, www.realestate.gov.cn.

Ghai, Y. *Hong Kong's New Constitutional Order: The Resumption of Chinese Sovereignty and the Basic Law.* Hong Kong: Hong Kong University Press, 1997.

Grace, G. "Education Is a Public Good: On the Need to Resist the Domination of Economic Science." In *Education and the Market Place,* edited by D. Bridges and T.H. McLaughlin. London: Falmer Press, 1994.

Grover, C. and J. Stewart. "Market Workfare: Social Security, Social Regulation and Competitiveness in the 1990s." *Journal of Social Policy,* 28, no. 1, 73–96.

Grover, C., and J. Stewart. "Modernizing Social Security? Labour and Its Welfare-to-work Strategy." *Social Policy & Administration* 34, no. 3 (September 2000): 235–52.

Gu, J., D.D. Lin, J.H. Chu, and H.Y. Zhu. "Zhongguo Weilao Fuwu Canye Zhengce Yanjiu" (Research on the Policy towards Service Industry for the Elderly in China). In *21 Shiji Shangbanye Zhongguo Laoling Wenti Duice Yanjiu* (Research on China's Policy Response to Population Aging Issues in the First Half of the 21 Century), edited by W. Zhang, et al. Beijing: Hualing Chubanshe, 2000.

Gui, S.X. "A Study on Long Term Care for Shanghai's Aging Population." Paper presented at the Eighth Annual Congress of Gerontology, Hong Kong Association of Gerontology, November 2000.

Gui, S.X. "Policy Analysis on Care of the Older People in Shanghai." In *Shanghai yu Xianggang Shehui Zhengce Bijiao Yanjiu* (Comparative Study on Social Policy in Shanghai and Hong Kong), edited by S.X. Gui and L. Wong, 330–354. Shanghai: Huadong Shifan Daxue Chubanshe, 2003 (in Chinese).

Gui, S.X. "Shanghai Shehui Zhengce de Fazhan yu Tiaozhan" (Shanghai's Social Policy Development and Challenge). In *Shanghai yu Xianggang Shehui Zhengce Bijiao Yanjiu* (Comparative Study on Social Policy in Shanghai and Hong Kong), edited by S.X. Gui and L. Wong. Shanghai: Huadong Shifan Daxue Chubanshe, 2003 (in Chinese).

Habermas, J. *The Legitimation Crisis.* Boston: Free Press, 1976.

Haila, A. "Why Is Shanghai Building a Giant Speculative Property Bubble?" *International Journal of Urban and Regional Research* 23, no. 3 (1999): 584–88.

Hao, K.M. *Twenty Years of Reforms in China's Educational System.* Zhengzhou: Zhongzhou Guji Chuban She, 1998 (in Chinese).

Harvard Team. *Improving Hong Kong's Health Care System: Why and for Whom?* Hong Kong: Printing Department, March 1999.

Hawkins, J.N. "Centralization, Decentralization, Recentralization: Educational Reform in China." *Journal of Educational Administration* 38 (2000): 442–54.

Hayhoe, R. *China's Universities 1895–1995: A Century of Cultural Conflicts.* New York: Garland Publishing, 1996.

Health and Welfare Branch (Hong Kong Government). *Towards Better Health: A Consultation Document.* Hong Kong: Government Printer, 1993.

Health and Welfare Bureau (Hong Kong Special Administrative Region Government). *Life-long Investment in Health: Consultation Document on Health Care Reform.* Hong Kong, December 2000.

Hirst, P., and G. Thompson. "Globalization and the Future of the Nation State." *Economy and Society* 24 (1995): 408–42.

Ho, L.S. "The Way to a Healthier Health Care System." *Policy Bulletin* 11 (May/June). Hong Kong: Hong Kong Policy Research Institute, 1999, pp. 11–12.

Hodge, P. "The Poor and the People of Quality: Social Policy in Hong Kong." *Hong Kong Journal of Social Work* 10, no. 2 (1976): 2–17.

Hodge, P. "The Politics of Welfare." In *The Common Welfare: Hong Kong's Social Services,* edited by J.F. Jones. Hong Kong: Chinese University Press, 1981.

Hollifield, J. *Immigrants, Markets and States: The Political Economy of Post-War Europe.* Cambridge, MA: Harvard University Press, 1992.

Holman, A. "UK Housing Finance: Past Changes, the Present Predicament, and Future Sustainability." In *Directions in Housing Policy: Towards Sustainable Housing Policies for the UK,* edited by P. Williams, 196–99. London: Paul Chapman Publishing, 1997.

Hong Kong Census and Statistics Department, *Annual Digest of Statistics.* Hong Kong: Government Printer, 2001.

Hong Kong Census and Statistics Department. *Annual Digest of Statistics.* Hong Kong: Government Printer, 2002.

Hong Kong Council of Social Service. *A Study on the Employment Situation of Middle and Old Aged Persons in Hong Kong.* Hong Kong: Hong Kong Council of Social Service Research Department, 2000.

Hong Kong Council of Social Service. *Research Report on the Problem of Unemployment among Middle-Aged to Older Workers in Hong Kong.* Hong Kong: Hong Kong Council of Social Service Research Department, 2000.

Hong Kong Council of Social Service. *Research Report on the Implementation of the Support for Self-Reliance Scheme.* Hong Kong: Hong Kong Council of Social Service Research Department, 2001 (in Chinese).

Hong Kong Council of Social Service. *Information on Services for the Elderly in Hong Kong—2002.* Hong Kong: Hong Kong Council of Social Service, 2002 (in Chinese).

Hong Kong Economic Journal, Hong Kong.

Hong Kong Government. *White Paper—Social Welfare into the 1980s.* Hong Kong: Government Press, 1979.

Hong Kong Government. *Estimates for the Year Ending 31 March 1986.* Vol. 1, *Expenditure.* Hong Kong: Government Printer, 1985.

Hong Kong Government. *White Paper—Social Welfare into the 1990s and Beyond.* Hong Kong: Government Press, 1991.

Hong Kong Government. *Report of the Working Group on Care of the Elderly.* Hong Kong: Government Press, 1994.

Hong Kong Government (HKG). *Building Hong Kong For a New Era.* Address by the Chief Executive The Honourable Tung Chee Hwa at the Provisional Legislative Council meeting on 8 October. Hong Kong: Hong Kong Government Printing Department, 1997.

Hong Kong Government (HKG). *From Adversity to Opportunity.* Address by the Chief Executive The Honourable Tung Chee Hwa at the Legislative Council meeting on 7 October. Hong Kong: Hong Kong Government Printing Department, 1998.

Hong Kong Government. *1999/2000 Government Budget*. Hong Kong: Hong Kong Government Printer, 1999.

Hong Kong Government (HKG). *Quality People, Quality Home: Positioning Hong Kong for the 21st Century*. Address by the Chief Executive The Honourable Tung Chee Hwa at the Legislative Council meeting on 6 October. Hong Kong: Hong Kong Government Printing Department, 1999.

Hong Kong Government. *Hong Kong Annual Report 1999*. Hong Kong: Government Printer, 2000.

Hong Kong Government (HKG). *Serving the Community, Sharing Common Goals*. Address by the Chief Executive The Honourable Tung Chee Hwa at the Legislative Council meeting on 11 October. Hong Kong: Hong Kong Government Printing Department, 2000.

Hong Kong Government Planning Department Web site: www.info.gov.hk/planning/index_e.htm. 1998 figures accessed on 5 July 2000.

Hong Kong Housing Authority. *1995/96 Annual Report*. Hong Kong: Hong Kong Housing Authority, 1996, 74.

Hong Kong Housing Authority. *Housing Authority Performance Indicators*, March 2001 edition and editions of earlier years.

Hong Kong Housing Authority Paper: HA34/2002, 21 June 2002.

Hong Kong Housing Authority Press Release, 29 August 2002.

Hong Kong Housing Society Web site: www.hkhs.com/, accessed in June 2000.

Hong Kong—Issues Facing an Aging Society. Hong Kong: Cosmos Books Ltd., 2002.

Hong Kong Mortgage Corporation Limited Web site: www.hkmc.com.hk/.

Hong Kong Policy Research Institute. *Health Care Reform: A Survey of the Views of Various Sectors*. Hong Kong: Hong Kong Policy Research Institute, 1999.

Hong Kong SAR Government. *Home for Hong Kong People into the 21st Century—A White Paper on Long Term Housing Strategy in Hong Kong*, 1998, paragraph 5.1.

Hong Kong Special Administrative Region Government. *Hong Kong in Figures* (2000). Government Web site: www.info.gov.hk/censtatd/eng/hkstat/hkinf/health, accessed on 2 November 2001.

Hood, C. "A Public Management for All Seasons." *Public Administration* 69 (1991): 3–19.

Hood, C. "The New Public Management in the 1980s: Variations on a Theme." *Accounting, Organizations and Society* 20 (1995): 93–109.

Hooyman, N.R., and H.A. Kiyak. *Social Gerontology: A Multidisciplinary Perspective* (6th ed.). Boston: Allyn and Bacon, 2002.

Hospital Authority, Hong Kong. Hospital Authority Web site: www.ha.org.hk, accessed 28 March 2001.

Hospital Authority, Hong Kong. Hospital Authority Web site: www.ha.org.hk, accessed 25 November 2002.

Hospital Authority, Hong Kong. *Overcoming Challenges for Better Health—Annual Plan 1999–2000*. Hong Kong: Hong Kong Hospital Authority, April 1999a.

Hospital Authority, Hong Kong. *Submission on Improving Hong Kong's Health Care System*. Hong Kong: Hong Kong Hospital Authority, August 1999b.

Hospital Authority, Hong Kong. *Rising to the Challenge: Annual Plan 2000–01*, Volume 1. Hong Kong: Hong Kong Hospital Authority, 2000.

Howe, C. "Foreword." In *Labour and the Failure of Reform in China*, edited by M. Korsez. London: Macmillan, 1992.

Howe, C., ed. *Shanghai: Revolution and Development in an Asian Metropolis*. Cambridge: Cambridge University Press, 1981.

Huang, T.J. "More Than 10 Billion Yuan Shanghai HPF Mortgage Loan to Individuals Were Made in 2001." *Shichang Bao* (Market Daily), 14 January 2002a.

Huang, T.J. "Strong Demand for Loan from Shanghai Housing Provident Fund Despite Limited Room for Additional Growth of Accumulated Fund." *Jingji Cankao Bao* (Reference News on Economy), 22 May 2002b.

Huang Z.G. "Progress and Extension of the Housing System Reform." *Housing for the New Era Conference Papers*, Shanghai-Hong Kong Housing Conference, Shanghai, 8–10 January 1997, 50–53.

Hughes, O.E. *Public Sector Reform in Hong Kong: Key Concepts, Progress-to-Date, Future Directions.* Hong Kong: Chinese University of Hong Kong Press, 1994.

Information Services Department. *Hong Kong 1998.* Hong Kong: The Government Printer, 1999.

Information Services Department, Hong Kong. "Public Health." *Hong Kong: The Facts,* (January 2000): www.info.gov.hk.

International Labour Organization. *Information Centres.* Geneva: International Labour Organization, 2000.

James, E. "Why Do Different Countries Choose a Different Public-Private Mix of Educational Services." *Journal of Human Resources* (June 1992): 571–92.

Jiang, Z.M. Speech at the Opening of the 15th Congress of the Chinese Communist Party, 12 September 1997.

Johnson, C.A. *MITI and the Japanese Miracle: The Growth of Industrial Policy, 1925–1975.* Stanford, CA: Stanford University Press, 1982.

Johnson, N. *Reconstructing the Welfare State.* Hempstead: Harvester Wheatsheaf, 1990.

Jones, P.W. "Globalization and Internationalism: Democratic Prospects for World Education." *Comparative Education* 34 (1998): 143–55.

Kam, P.K., and Y.H. Kwan. "The Rights and Benefits of Elderly—The Sad Song of Hong Kong." *Hong Kong Journal of Gerontology* 10, no. 1 (1996): 51–57.

Kang, Y. *Interpreting Shanghai: 1990–2000.* Shanghai: Shanghai People's Press, 2001 (in Chinese).

Kaple, D. *Dream of a Red Factory.* New York: Oxford University Press, 1994.

Ketiju. *Shanghaishi Laoling Baozhang Tixi ji qi Yunxing Jizhi Yanjiu* (Research on Shanghai's Old Age Security Framework and Its Operating System). Shanghai: Shanghai Zeji Jixu Wenxian Chubanshe, 1998.

Ko, K. "Property Transactions Down 52pc in First Half." *South China Morning Post,* 4 July 1998.

Kornai, J. *Economics of Shortage.* Amsterdam: North-Holland, 1980.

Kwan, Y.H. "The Old, the Aged, and the Elderly." In *The Other Hong Kong Report,* edited by J.Y.S. Cheng, 431–53. Hong Kong: Chinese University Press, 1997.

Kwan, Y.H. *All about Elderly Welfare.* Hong Kong: Cosmos Books Ltd., 2000 (in Chinese).

Kwan, Y.H., ed. *Aging Hong Kong—Issues Facing an Aging Society.* Hong Kong: Cosmos Books, Ltd., 2000.

Kwan, Y.H., and S.C. Chan. "The Psycho-social Care in Institutions for the Elderly in Beijing, Shanghai, Taiwan and Hong Kong." In *Conference Proceedings of Asia-Pacific Regional Conference for the International Year of Older Persons,* 26–29 April 1999a, Hong Kong, 221–33.

Kwan, Y.H., and S.C. Chan. "The Challenge of Health Care in Institutions for the Elderly in Urban China." In *China in the Reform Era,* edited by X.W. Zang, 81–101. New York: Nova Science Publishers, 1999b.

Kwan, Y.H., S.X. Gui, and M.L. Chong. "Policy Analysis on Care of the Older People—Comparison Between Shanghai and Hong Kong." In *Shanghai yu Xianggang Shehui Zhengce Bijiao Yanyiu* (Comparative Study on Social Policy in Shanghai and Hong Kong), edited by S.X. Gui and L. Wong, 376–88. Shanghai: Huadong Shifan Daxue Chubanshe, 2003.

Lam, C. "Allocation of New Welfare Service Units." Paper presented at the Welfare Panel of the Legislative Council, HKSAR. Hong Kong: Social Welfare Department, 5 March 2001.

Lam, C. "Director of Social Welfare's Speech at the Tenth Anniversary of the Bachelor of Social Work Program," City University of Hong Kong, Hong Kong, 2 June 2001.

Lam, K.C., and P.W. Liu. *Immigration and the Economy of Hong Kong*. Hong Kong: City University of Hong Kong Press, 1998.

Lau, K.Y. "The Implications of the Inadequate Supply and Inequitable Public Sector Housing Allocation Policies." In *Commentary on Housing Policies in Hong Kong*, edited by P.K. Kam et al., 62–66. Hong Kong: Joint Publishing, 1996.

Lau, K.Y. "Safeguarding the Rational Allocation of Public Housing Resources in Hong Kong: A Social Policy Discussion." *Asia Pacific Journal of Social Work* 7, no. 2 (1997): 97–129.

Lau, K.Y. "Housing Privatization: Some Observations on the Disengagement and Delimitation Strategies." In *Housing Express*, 11–16. Hong Kong: Chartered Institute of Housing, November 2001.

Lau, S.K. *Society and Politics in Hong Kong*. Hong Kong: Chinese University Press, 1982.

Leading Group of the Project on Shanghai into the 21st Century, ed. *Shanghai into the Twenty-First Century: Strategic Research on Shanghai's Economy and Social Progress from 1996 to 2010*. Shanghai: Shanghai People's Press, 1995.

Lee, G.O.M. "Hong Kong." In *Labour Management Cooperation: From Labour Disputes to Co-operation*. Tokyo: Asian Productivity Organization, 1996.

Lee, H.Y. "Xiagang, The Chinese Style of Laying Off Workers." *Asian Survey* 40, no. 6 (2000): 914–37.

Lee, J. "Xinyimin Wenti yu Shehui Zhengce" (The Problem of Immigrants and Social Policy). In *Xin Shehui Zhengce* (New Social Policy), edited by J. Lee, S. Chiu, L.C. Leung, and K.W. Chan, 269–81. Hong Kong: Chinese University Press, 1998.

Lee, J., and A. Cheung. *Public Sector Reform in Hong Kong*. Hong Kong: Chinese University of Hong Kong Press, 1995.

Lee, K.C., W.S. Chiu, L.C. Leung, and K.W. Chan. *Xin Shehui Zhengce* (New Social Policy). Hong Kong: Chinese University Press, 1999.

Lee, L. Ou-fan. *Shanghai Mo Deng: Yizhong Xin Dushi Wenhua zai Zhongguo 1930–1945* (Shanghai Modern: A New Type of Urban Culture in China 1930–1945). Beijing: Beijing Daxue Chubanshe, 2001.

Legislative Council, Hong Kong. *Official Record of Proceedings*, 5 May 1999, accessed on Web site: www.legco.gov.hk, 1 November 2001.

Le Grand, J., and R. Robinson. *Privatization and the Welfare State*. London: George Allen and Unwin, 1985.

Le Grand, J., and W. Bartlett, eds. *Quasi-Markets and Social Policy*. London: Macmillan, 1993.

Leung, J. "The Emergence of Unemployment Insurance in China." *Canadian Review of Social Policy* 38 (fall 1996a): 5–19.

Leung, J. *Family Support for the Elderly in China: Continuity and Change*. Monograph Series "Social Welfare in China," no. 5. Hong Kong: Department of Social Work and Social Administration, University of Hong Kong, 1996b.

Levy, D. *Higher Education and the State in Latin America: Private Challenges to Public Dominance*. Chicago: University of Chicago Press, 1986.

Li, C.F. "A Study of the Unemployment Problem and Government's Prescriptions in Hong Kong." *China Development Briefing* 1 (1999): 19–34.

Liu, P., W.H. Liu, B.J. Xiong, and Y. Cheng. "21 Shiji Shangbanye Woguo Laolinghua Wenti de Duice Yanjiu" (Research on the Policy Response to Old Age Issues in Our Country in the First Half of the 21st Century). In *21 Shiji Shangbanye Zhongguo Laolinghua Wenti Duice Yanjiu* (Research on China's Policy Response to Population

Aging Issues in the First Half of the 21st Century), edited by W.F. Zhang, et al., 18–50. Beijing: Hualing Chubanshe, 2000.

Liu, Y.M., and X.F. Shang. "Shanghaishi Jiuwunian he Jiuqinian Liudong Renkou de Shouru Yinsu Bejiao Fenxi" (Comparative Analysis of Income Factor of Floating Population in Shanghai 1995 and 1997). *Renkou* 4 (1998): 34–44.

Liu, Z.F. Speech presented at the Tenth Anniversary of the Shanghai Housing Provident Fund Commemorative Meeting, 25 April 2001.

Lu, H.L. "Economic Reforms and Diversification of Social Life." In *A Report of Social Development in Shanghai 2001*, edited by J.Z. Yin. Shanghai: Academy of Social Science, 2001.

Lu, X., and E.J. Perry, eds. *Danwei: The Changing Chinese Workplace in Historical and Comparative Perspective*. New York: M.E. Sharpe, 1997.

MacPherson, R.J.S. *Educative Accountability*. Oxford: Pergamon, 1996.

Mannheim, K. *Ideology and Utopia: An Introduction to the Sociology of Knowledge*. London: Routledge, 1954.

Mason, R. *Globalising Education: Trends and Applications*. London: Routledge, 1998.

McLaughlin, E. "Hong Kong: A Residual Welfare Regime." In *Comparing Welfare States: Britain in International Context*, edited by A. Cochrane and J. Clarke, 105–50. London: Sage, 1993.

MDR. *Survey of Housing Aspiration of Households—Final Report*, 2000. Prepared for the Planning Department of the Hong Kong Special Administrative Region Government, Table 7, 23.

Miners, N. *The Government and Politics of Hong Kong*. Hong Kong: Oxford University Press, 1991.

Ming Pao, Hong Kong, various issues.

Ming Pao. "Every Year Four Thousand Elderly on the Waiting List to Old Age Homes Die while Waiting," 6 June 2000, A7.

Ming Pao. "Hong Kong People: The World's Third in Longevity," 25 July 2002, A4.

Ministry of Civil Affairs. *1992 White Paper: A Report on the Development of China's Social Welfare Services*. Beijing: Ministry of Civil Affairs, 1993.

Ministry of Civil Affairs. *Report on Annual Work*. Shanghai: Shanghai Department of Civil Affairs, 1999.

Ministry of Health, People's Republic of China. *Research on National Health Services 1999*. Beijing: Ministry of Health, 1999.

Ministry of Information and the Arts, Singapore. *Singapore: Facts and Pictures 2000*. Singapore: Ministry of Information and Arts, 2000.

Ministry of Labor. *Annual Departmental Report*. Shanghai: Ministry of Labor, 1997.

Mitzman, A. *The Iron Cage: An Historical Interpretation of Max Weber*. New York: Grosset and Dunlap, 1971.

Mok, K.H. "Marketization and Quasi-Marketization: Educational Development in Post-Mao China." *International Review of Education* 43 (1997): 1–21.

Mok, K.H. "Education and the Market Place in Hong Kong and Mainland China." *Higher Education* 37 (1999): 133–58.

Mok, K.H. "Marketizing Higher Education in Post-Mao China." *International Journal of Educational Development* 20 (2000): 109–36.

Mok, K.H., and D. Chan. "The Emergence of Private Education in the Pearl River Delta: Implications for Social Development." In *The Economic and Social Developments in South China*, edited by S. MacPherson and J. Cheng. Cheltenham: Edward Elgar, 1996.

Mok, K.H., and D. Chan. "Privatization or Quasi-Marketization." In *Higher Education in Post-Mao China*, edited by M. Agelasto and B. Adamson. Hong Kong: University of Hong Kong Press, 1998.

Mok, K.H., and D. Chan. "Educational Development and the Socialist Market in Guangdong." *Asia Pacific Journal of Education* 21 (2001): 1–18.
Mok, K.H., and W.P. Huen. "Unemployment in China: Policy Responses, Coping Strategies
and Policy Implications." *China Development Briefing* 1 (1999): 3–18.
Mok, K.H., and K.Y. Wat. "Merging of the Public and Private Boundary: Education and
the Market Place in China." *International Journal of Educational Development* 18
(1998): 255–67.
Morris, J. *Hong Kong: Epilogue to Empire.* London and New York: Penguin Books,
1997.
Murphy, R. *Shanghai: Key to Modern China.* Cambridge: Harvard University Press, 1953.
Nanfang Fangdichan (Southern Housing and Real Estate). January 2000, p. 30.
Ng, S.H., and G.O.M. Lee. "Hong Kong Labor Market in the Aftermath of the Crisis:
Implications for Foreign Workers." *Asian and Pacific Migration Journal* 7, no. 2–3
(1998): 171–85.
Ngan, M.H., M.F. Leung, Y.H. Kwan, W.T. Yeung, and M.L. Chong. *A Study of the Long-
Term Care Needs, Patterns and Impact of the Elderly in Hong Kong.* Hong Kong:
Department of Applied Social Studies, City University of Hong Kong, 1996.
Ngan, R. "Social Security and Poverty." In *Social Work in Hong Kong,* edited by I. Chi
and S.K. Cheung. Hong Kong: Hong Kong Social Workers' Association, 1996.
Ngan, R. "Pension Reform in China: Development and Critique." *China Development Brief-
ing* 3 (1999): 6–19.
Ngan, R., and E. Leung. "Long-term Care in Hong Kong: The Myth of Social Support and
Integration." In *Elderly Chinese in Pacific Rim Countries,* edited by I. Chi, N.L. Chap-
pell, and J. Lubben, 15–34. Hong Kong: Hong Kong University Press, 2001.
Ngok, K.L., and D. Chan. "Shanghai Education into the 21st Century: A Quest for Learning
Society." *Chulalongkorn Educational Review* 6 (2000a): 1–14.
Ngok, K.L., and D. Chan. "Lifelong Learning and Learning City: Shanghai Education Re-
form and Development Towards the 21st Century." *Educational Policy Forum* 3
(2000b): 80–98 (in Chinese).
Ngok, K.L., and R. Yep. "Chinese Immigrants in Hong Kong: Policy and Problems." Un-
published manuscript, Department of Public and Social Administration, City University
of Hong Kong, Hong Kong, 2000.
Offe, C. "Reflection on the Institutional Self-Transformation of Movement Politics: A Ten-
tative Stage Model." In *Challenging the Political Order: New Social and Political
Movements in Western Democracies,* edited by R.J. Dalton and M. Koehler. New York:
Oxford University Press, 1990.
Osborne, D., and T. Gabler. *Reinventing Government: How the Entrepreneurial Spirit Is
Transforming the Public Sector.* Reading, MA: Addison Wesley, 1992.
Panel on Health Services, Hong Kong Legislative Council. *Minutes of Meeting held on 12
March 2001,* LC Paper, no. CB(2)1224/00–01, Hong Kong, 2001.
Pang, N.S.K. "The School Management Reform in Hong Kong: Restructuring the Account-
ability System." *Education Journal* 25 (1997): 25–44.
Pang, N.S.K. "Should Quality School Education be a Kaizen (Improvement) or an Innova-
tion?" *International Journal of Educational Reform* 7 (1998): 1–11.
Peng, R.C., R.H. Cai, and C.M. Zhou. *Zhongguo gaige chuanshu: Yilao weisheng tizhi gaige
juan 1978–91* (Medical and Health System Reform, China Reform Collection Series
1978–91). Dalian, China: Dalian Press, 1992.
Phillipson, C., and N. Thompson. "The Social Construction of Old Age." In *Developing
Services for Older People and Their Families,* edited by Rosemary Bland. London:
Jessica Kingsley Publishers, 1996.

Pollitt, C. *Management and the Public Service: The Anglo-American Experience*. Oxford: Blackwell, 1990.

Project Team on Social Development Issues. *Shanghai Kuashiji Shehuifazhan Wenti Sikao* (Reflections on Shanghai's Cross-Century Social Development Issues). Shanghai: Shanghai Social Science Academy Press, 1997.

Provisional Hospital Authority, Hong Kong. *Report of the Provisional Hospital Authority*. Hong Kong, December 1989.

Pusey, M. *Economic Rationalism in Canberra*. Melbourne: Cambridge University Press, 1991.

Pye, L.W. "Foreword." In *Shanghai: Revolution and Development in an Asian Metropolis*, edited by C. Howe, xi–xvi. Cambridge: Cambridge University Press, 1981.

Qiao, J. "Report of the Condition of the Chinese Workers in 1998." In *China in 1999: Analysis and Forecast of Social Conditions*, edited by X. Ru, X.Y. Lu, and T.L. Shan. Beijing: Social Science Publisher, 1999 (in Chinese).

Rees, S. "The Fraud and the Fiction." In *The Human Costs of Managerialism*, edited by S. Rees and G. Rodley. Sydney, Australia: Pluto Press, 1995.

Rosenau, J., ed. *Governance without Government: Order and Change in World Politics*. Cambridge: Cambridge University Press, 1992.

Saunders, P., and X.Y. Shang. "Social Security Reform in China's Transition to a Market Economy." *Social Policy & Administration* 35, no. 3 (July 2001): 274–89.

Schattschneider, E.E. *The Semi-Sovereign People*. Hinsdale, IL: Dreyden Press, 1975.

Scott, I. *Political Change and the Crisis of Legitimacy*. Hong Kong: Oxford University Press, 1989.

Scott, W.D. *The Delivery of Medical Services in Hospitals: A Report for the Hong Kong Government*. Hong Kong, December 1985.

Shanghai Academy of Social Science Housing and Real Estate Research Institute. *A Research Report on Position and Prospect of Shanghai Housing and Real Estate Market*. Shanghai: Shanghai Academy of Social Science, 2001.

Shanghai City Housing Development Bureau. "Speed Up Modernization Steps on Housing Industry to Promote Sustainable Housing Development in Shanghai." www.cin.gov.cn/fdc/exp/2001102307.htm.

Shanghai City Housing System Reform Office. "An Effective Attempt in Marketisation of Public Sale Flats." *Zhuzhai yu Fangdichan* (Housing and Real Estate) (April 1998): 25–27.

Shanghai Economy Yearbook 2000, p. 352.

Shanghai Economy Yearbook 2001, p. 277.

Shanghai Encyclopaedia. Shanghai: Shanghai Scientific Press, 1999.

Shanghai Housing and Land Resource Bureau. "Building the Housing Consumption System for General Public." www.cin.gov.cn/fdc/exp/2000092902.htm.

Shanghai Jingji Xuehui. *Kaifa, Kaifo, Kaifang—Shanghai Jingji Fazhan Zhonglun* (Development, Port Founding and Opening—Comprehensive Commentary on the Economic Development in Shanghai). Beijing: Zhongguo Guangbo Dianshi Chubanshe, 1992.

Shanghai Municipal Commission of Education. *Shanghai Education Yearbook 1998*. Shanghai: Shanghai Education Press, 1998.

Shanghai Municipal Commission of Education. *Collection of Documents of 1999 Shanghai Municipal Working Meeting on Education*. Shanghai: Century Publishing Group and Education Press, 2000.

Shanghai Municipal Committee on Aging and Shanghai Research Center on Aging. *The Report of Population Aging in Shanghai—2001*. Shanghai: Shanghai Research Centre on Aging, 2002 (in Chinese).

Shanghai Municipal Government. "Fully Implement Quality Education, Cultivate Excellent Talents and High Quality Laborers for Shanghai's Development in the Next Century."

In *Deepen Education Reform and Push Forward Quality Education: Proceedings of the Third National Education Working Meeting*, edited by Ministry of Education. Beijing: Higher Education Press, 1999.

Shanghai Municipal Government. *Shanghai Renmin Zhengfuling Dijiuhao Shanghaishi Chengzhen Zhigong Jiben Yiliao Baoxian Banfa* (Municipal Government Order, no. 99: The Shanghai Municipality's City and Township Employees' Basic Medical Security Measures), 20 October 2000, accessed on Web site www.sh.gov.cn, 21 November 2001.

Shanghai Municipal Health Bureau. *Shanghaishi 1949–80 Nian Weisheng Tongji Ziliao* (Shanghai City Health Statistics, 1949–80). Shanghai, 1982 (in Chinese, not for open circulation).

Shanghai Municipal Health Bureau. *Shanghai chengshi yiliao weisheng fuwu yanjiu* (Shanghai: Research on Shanghai City Medical and Health Services), 1987 (in Chinese, not for open circulation).

Shanghai Municipal Health Bureau. *Shanghai Health Yearbook*. Shanghai: Shanghai Science and Technology Documents Press, 1991–2000 (various years) (in Chinese).

Shanghai Municipal Statistics Bureau. *Shanghai Statistics Yearbook*. Shanghai: China Statistics Press, 1993, 1998, 2000 (in Chinese).

Shanghai Municipal Statistics Bureau. *Shanghai Statistics Yearbook [Shanghai Tongji Nianjian]*. Shanghai: China Statistics Press, 2001 (in Chinese).

Shanghai People's Congress Standing Committee. *Shanghaishi Wailai Liudong Renyuan Guanli Tiaoli* (Regulations of Management of Floating Personnel in Shanghai). Shanghai, 1996.

Shanghai Personnel Bureau. *Shanghaishi Wailai Renkou Gongzuo Jizhuzheng Shishi Banfa* (Method of Implementation of the Work Temporary Resident Permit Scheme in Shanghai), Shanghai, 1998.

Shanghai Research Center on Aging. *Information on Older Population in Shanghai—2001*. Shanghai: Shanghai Research Centre on Aging, 2001 (in Chinese).

Shanghai Residential Housing 1949–1990 Editorial Board. *Shanghai Residential Housing 1949–1990*. Shanghai: Shanghai Kexue Puji Chubanshe, 1993, 16, 33, 134.

Shanghaishi Laoling Kexue Yanjiu Zhongxin, ed. *Shanghaishi 2000 Nian Laonian Renkou Xinxi* (Shanghai City Aged Population Information 2000). Shanghai: Shanghai Laoling Kexue Yanjiu Zhongxin, 2001.

Shanghai Statistical Bureau. *Shanghai Statistical Yearbook*. Shanghai: Shanghai Statistical Bureau, 1999.

Shanghai Statistical Bureau. *Shanghai Statistical Yearbook*. Shanghai: Shanghai Statistical Bureau, 2001.

Shanghai Statistical Yearbook 2001, Table 20.56 (Internet edition).

Shen, A., and Y. Yang. "Meeting the Living Requirements and Improving the Quality of Life." In *A Report of Social Development in Shanghai 2001*, edited by J.Z. Yin, 50–51. Shanghai: Shanghai Shehui Kexueyuan Chubanshe, 2001.

Shen, Z.X. et al. *Shanghaishi Renkou Laolinghua Baogaoshu* (Report on Population Aging in Shanghai). Shanghai: Shanghai Laoling Weiyuanhui, Shanghaishi Laoling Kexue Yanjiu Zhongxian, 2000.

Shen Z.X. et al., eds. *Shanghaishi Renkou Laolinghua Baogaoshu* (Shanghai City Population Aging Report). Shanghai: Shanghaishi Laoling Weiyuanhui, Shanghaishi Laoling Kexue Yanjiu Zhongxin, 2000.

Shenkar, O., and M.A. Von Glinow. "Paradoxes of Organizational Theory and Research: Using the Case of China to Illustrate National Contingency." *Management Science* 40, no. 1 (1994): 56–71.

Simon, J. *The Economic Consequences of Immigration*. Oxford: Blackwell, 1989.

Sinn, E. *Power and Charity: The Early History of the Tung Wah Hospital, Hong Kong.* Hong Kong: Oxford University Press, 1989.

Slaughter, S., and L. Leslie. *Academic Capitalism.* Baltimore, MD: Johns Hopkins University Press, 1997.

Smart, A., and J. Lee. "Financialization and the Role of Real Estate in Hong Kong's Regime of Accumulation." *Economic Geography* 78 (2003), forthcoming.

Social Science Research Centre. *Pop Express.* University of Hong Kong, Special Release of 25 June 1998a.

Social Science Research Centre. *Pop Express.* University of Hong Kong, Special Release of 30 June 1998b.

Social Welfare Department. *Support for Self-reliance: A Review of CSSA. Review Report.* Hong Kong: Government Printer, 1998.

Social Welfare Department. *Annual Departmental Report.* Hong Kong: Government Printer, 2000.

Social Welfare Department. *Evaluation Report on Support for Self-Reliance Scheme.* Presented to the Health & Welfare Panel, Legislative Council, Hong Kong, 2001.

Social Welfare Department. Homepage of Social Welfare Department. Retrieved on March 30, 2001. www.info.gov.hk/swd/html_tc/index.html.

Solinger, D. *Contesting Citizenship in Urban China, Peasant Migrants, the State and the Logic of the Market.* Berkeley: University of California Press, 1999.

South China Morning Post, Hong Kong, various issues.

Spring, J. *Education and the Rise of the Global Economy.* Mahwah, NJ: Lawrence Erlbaum Associates, 1998.

State Council. *Zhongguo Gongmin Yin Sishi Wanglai Xianggang Diqu Huozhe Aomen Diqu de Zanxing Guanli Banfa,* (Provisional Regulation on the Management of Chinese Citizen's Migrating to Hong Kong or Macau for Personal Reasons), 1986.

State Education Commission (SEC) Policies and Law Department. *Law and Regulation on Basic Education of the People's Republic of China.* Beijing: Beijing Normal University, 1993 (in Chinese).

Statistical Reports on Shanghai Municipal National Economic and Social Development 2000 and 2001. http://tjj.sh.gov.cn/tjgb/tjgb.htm and http://tjj.sh.gov.cn/tjgb/gb2001.htm.

Su, S.X., and J.Q. Chen. *Fifteen-year of Education Reform in Shanghai.* Shanghai: Shanghai Academy of Social Sciences Press, 1994.

Sun, C.S. et al., eds. *Qiji shi Ruhe Chuangzao Chulai de: Guanyu Shanghaishi Zaijiuye Gongcheng de Yanjiu Baogao* (The Making of a Miracle: Research Report on the Re-Employment Project in Shanghai). Shanghai: Fudan University Press, 1998.

Sun, G. "Health Care Administration in China." In *Public Policy in China,* edited by S.S. Nagel, 53–62. Westport, CT: Greenwood Press, 1992.

Sun, Z.F. *Shanghaishi Liudong Renkou Zeju Xingwei ji qi Jujuqu de Xingcheng Jizhi* (The Residential Pattern of Mobile Population in Shanghai). Unpublished Master's thesis, Shanghai, East China Normal University, 1999.

Sweeting, A. *A Phoenix Transformed: The Reconstruction of Education in Post-War Hong Kong.* Hong Kong: Oxford University Press, 1993.

Ta Kung Pao, 6 November 2001.

Tam, W.M., and Y.C. Cheng. "A Multi-Model Analysis of Systematic School Quality Improvement Plans: The Case of Hong Kong." Paper presented at the 14th Annual Conference of the Hong Kong Educational Research Association, Hong Kong, 14–16 November 1997.

Tang, J. "Zhongguo Shehui Jiuzhu Zhidu Gaige Baogao." In *Jichu Zhenghe de Shehui Baozhang Tixi* (Social Security System under Foundational Reorganization), edited by T.K. Jing. Beijing: Huaxia Chubanshe, 2001.

Tang, K.L. *Colonial State and Social Policy: Social Welfare Development in Hong Kong 1842–1997*. Lanham, MD: University Press of America, 1998.

Tang, K.L., and R. Ngan. "China: Developmentalism and Social Security." *International Journal of Social Welfare* 10 (2001): 251–57.

Task Force on the Strategy of Education Development in Shanghai. *Research Report on Strategic Development of Shanghai's Education*. Shanghai: East China Normal University Press, 1989.

Taylor, S., F. Rizri, B. Lingard, and M. Henry, eds. *Educational Policy and the Politics of Change*. London: Routledge, 1997.

Thong, K.L. "Medical and Health Systems with Special Reference to Hong Kong." *Journal of The Hong Kong Society of Community Medicine* (The 15th Digby Memorial Lecture special issue) 17, no. 2 (September 1987): 1–40.

Tian, X., J. Zhan, and T. Zhao. "Renkou Laolinghua yu Jingji Fazhan" (Population Aging and Economic Development). In *21 Shiji Shangbanye Zhongguo Laoling Wenti Duice Yanjiu* (China's Policy Response to the Old Age Issue in the First Half of the 21st Century), edited by W.F. Zhang et al., 78–112. Beijing: Hualing Chubanshe, 2000.

Tilak, J. "The Privatization of Higher Education." *Prospects* 21, 1991, 227–39.

Tin Tin Daily, Hong Kong, various issues.

Tsang, D. *The 1997–1998 Budget*. Speech by the Financial Secretary at the Legislative Council, March 1998.

Tsang, D. *Scaling New Heights*. Speech by the Financial Secretary moving the Second Reading of the Appropriation Bill 2000 at the Legislative Council, Hong Kong, 8 March 2000.

Tsang, D. *The 2001–2002 Budget of the Hong Kong Special Administrative Region Government*. Retrieved on 31 March 2001. www.info.gov.hk/budget01–02/bgtshf.html.

Tung, C.H. *A Future of Excellence and Prosperity for All*. Speech by the Chief Executive at the ceremony to celebrate the establishment of the Hong Kong Special Administrative Region of the People's Republic of China, Hong Kong, 1 July 1997.

Tung, C.H. *Chief Executive's Policy Address 2000*. Hong Kong: Government Printer, 2000.

Tyler, W. "The Organizational Structure of the School." *Annual Review of Sociology* 15 (1985): 49–73.

Unger, B. "Limits of Convergence and Globalization." In *The Political Economy of Globalization*, edited by S.D. Gupta, 99–127. Boston: Kluwer, 1997.

United Nations. *Human Development Report 1999*. New York: Oxford University Press, 1999.

United Nations. Department of Economics and Social Affairs. *World Population Prospect: Estimates and Projections as Assessed in 1982*. New York: United Nations, Department of Economics and Social Affairs, 1982.

U.S. Bureau of the Census. Department of Commerce. *Global Aging into the 21st Century*. Information Sheet. Washington: U.S. Bureau of the Census, 1996.

U.S. Bureau of the Census. International Database. *Midyear Population, by Age and Sex*. Retrieved on 31 March 2001. www.census.gov/cgi.bin/ipc/idbagg.

Walker, A. "Towards a Political Economy of Old Age." *Aging and Society* 1 (1981): 73–94.

Walker, A., and Walker. *The Caring Gap*. London: Local Government, 1985.

Walsh, K. *Public Services and Market Mechanisms*. London: Macmillan, 1995.

Wan, C.C., ed. *A Report of Economic Development in Shanghai 2000*. Shanghai: Shanghai Academy of Social Sciences Publisher, 2000 (in Chinese).

Wan, D.N., Y.Y. Zhang, and B.G. Chen. "Research on the Contribution Percentage to Shanghai Economic Development Due to Residential Consumption." *Chinese Real Estate Studies* 1 (2000): 81–96.

Wang, C., ed. *The General Direction: The Political Agenda of the New Chinese Government into the Next Century*. Beijing: Yehai Press, 1998.

Wang, J. "An Exploratory Study on the Residential Housing Price to Income Ratio and Rent to Income Ratio in Shanghai." *Chinese Real Estate Studies* 1 (2001): 52–53.

Wang, W.D. *Jiushi Niandai Shanghai Liudong Renkou* (Population Mobility in Shanghai in the 1990s). Shanghai: East China Normal University Press, 1995.

Wei, B.P.T. *Shanghai: Crucible of Modern China.* Hong Kong: Oxford University Press, 1987.

Welch, A. *Australian Education: Reform or Crisis?* Sydney, Australia: Allen & Unwin, 1996.

Welch, A. "The Cult of Efficiency on Education: Comparative Reflections on the Reality and the Rhetoric." *Comparative Education* 34 (1998): 157–75.

Wesley-Smith, P., ed. *Hong Kong's Transition: Problems and Prospects.* Hong Kong: Faculty of Law, University of Hong Kong, 1993.

Wesley-Smith, P., and A.H.Y. Chen, eds. *The Basic Law and Hong Kong's Future.* Hong Kong: Butterworths, 1988.

Westwood, R.I., and S.M. Leung. "Work under the Reforms: The Experience and Meaning of Work in a Time of Transition." In *China Review 1996,* edited by M. Brosseau, S. Pepper, and S.K. Tsang. Hong Kong: Chinese University Press, 1996.

White, B.S. *Turbans and Traders.* Hong Kong: Oxford University Press, 1994.

White, G. "Labour Market Reform in Chinese Industry." In *Management Reforms in China,* edited by Malcolm Warner. New York: St. Martin's Press, 1987.

White, L.T. "Shanghai-Suburb Relationship 1949–1966." In *Shanghai: Revolution and Development in an Asian Metropolis,* edited by C. Howe, 241–68. Cambridge: Cambridge University Press, 1981.

White, L.T. *Shanghai Shanghaied? Uneven Taxes in Reform China.* Hong Kong: Center of Asian Studies, University of Hong Kong, 1991.

White, L.T. "Migration and Politics on the Shanghai Delta." *Issues and Studies* 30, no. 9 (1994): 63–94.

White, L.T. "Shanghai's 'Horizontal Liaisons' and Population Control." In *Shanghai: Transformation and Modernization under China's Open Door Policy,* edited by Y.M. Yeung and Y.W. Sung. Hong Kong: Chinese University Press, 1996.

White, L.T. *Unstately Power: Local Causes of China's Economic Reform.* New York: M.E. Sharpe, 1998.

Whitty, G. "Marketization, the State and the Re-formation of the Teaching Profession." In *Education, Culture, Economy and Society,* edited by A.H. Halsey et al. Oxford: Oxford University Press, 1997.

Wilson, S., and A.V. Adams. *Self-Employment for the Unemployed: Experience in OECD and Transitional Economies.* Washington, DC: World Bank, 1994.

Wong, L. "Privatization of Social Welfare in Post-Mao China." *Asian Survey* 34, no. 4 (1994): 307–25.

Wong, L. *Marginalization and Social Welfare in China.* London and New York: Routledge and LSE, 1998.

Wong, L. "Xianggang Shehui Fuli de Tiaozhan ji qi yu Shanghai de Bijiao" (Hong Kong's Social Welfare Challenge and Its Comparison with Shanghai). *Shanghai Laoling Kexue* 1–2 (2001): 62–68.

Wong, L., and W.P. Huen. "Reforming the Household Registration System: A Preliminary Glimpse of the Blue Chop Household Registration System in Shanghai and Shenzhen." *International Migration Review* 32 (winter 1998): 974–94.

Wong, L., and N. Flynn, eds. *The Market in Chinese Social Policy.* New York: Palgrave, 2001.

Wong, S.L. *Emigrant Entrepreneurs—Shanghai Industrialists in Hong Kong.* Hong Kong: Oxford University Press, 1988.

Wong, S.L. "The Entrepreneurial Spirit: Shanghai and Hong Kong Compared." In *Shanghai: Transformation and Modernization under China's Open Policy,* edited by Y.M. Yeung and Y.W. Sung. Hong Kong: Chinese University Press, 1996.

Wong, S.L., and T. Maruya, eds. *Hong Kong Economy and Society: The New Era.* Hong Kong: Center of Asian Studies, University of Hong Kong, 1998.

Working Group. "Review of Shanghai Social Security for the Aged: System and Operation." In *Review of Shanghai Social Security for the Aged: System and Operation,* 64. Shanghai: Shanghai Scientific Publication Publisher, 1998.

Working Party on Primary Health Care, Hong Kong. *Health for All: The Way Ahead.* Hong Kong: Government Printer, 1990.

World Bank. *Report on China.* Washington, DC: World Bank, 1984.

World Bank. *Poverty Reduction and the World Bank: Progress and Challenges in the 1990s.* World Development Bank Publication, Geneva (also in *China Development Briefing,* Issue 3, October 1996, p. 22).

World Bank. *Vietnam: Education Financing Sector Study.* Washington, DC: World Bank, 1996.

World Bank. *Financing Health Care: Issues and Options for China,* China 2020 Series. Washington, DC: World Bank, 1997.

World Bank. *Education in Sub-Saharan Africa: Policies for Adjustment, Revitalisation and Expansion.* Washington, DC: World Bank, 1998a.

World Bank. *The Philippines: Education Sector Study,* Vol. 2. Washington, DC: East Asia and Pacific Regional Office, Country Development I, World Bank, 1998b.

World Health Organisation. *Population Aging–A Public Health Challenge.* Geneva: WHO, Health Communication and Public Relations, 1999. Fact Sheet, No. 135.

Wu, D. "Poverty and Welfare Relief Policies in Shanghai." Paper presented at the Conference on Social Policy Development in Shanghai and Hong Kong, City of University of Hong Kong, March 2000.

Wu, Z.T. "Housing Provident Fund's Full Implementation for Realisation of Home Owning." *Housing for the New Era Conference Papers,* pp. 54–55. Shanghai-Hong Kong Housing Conference, Shanghai, 8–10 January 1997.

Xie, J.J. Concluding Speech presented at the National Conference on Management of Housing Provident Fund and at the National Seminar on Dwellings and Real Estate Information, Shanghai, 30 October 2001.

Xinmin Wanbao (New People's Evening Post). Shanghai, 3 July 1996 (in Chinese).

Xu, H.T. "Mayor of Shanghai's Speech delivered at the 140 Anniversary Dinner of the Hong Kong General Chamber of Commerce," *Tai Kung Daily News,* 12 May 2001, A6.

Yang, T. *Shehui Fuli Shehuihua: Shanghai yu Xianggang Shehui Fulie Tixi Bijiao* (Social Welfare Socialization: Comparative Study on the Social Welfare Systems of Shanghai and Hong Kong). Beijing: Huaxia Chubanshe, 2001.

Yao, X.T. *Shanghai Xianggang Bijiao Yanjiu* (Shanghai and Hong Kong Comparative Study). Shanghai: Shanghai Renmin Chubanshe, 2001.

Yao, X.T., ed. *Shanghai Xianggang Bijiao Yanjiu* (Shanghai and Hong Kong Comparative Study). Shanghai: Shanghai Renmin Chubanshe, 1990.

Yee, A.H., and T.G. Lim. "Education Supply and Demand in East Asia: Private Higher Education." In *East Asian Higher Education: Traditions and Transformations,* edited by A.H. Yee. Oxford: Pergamon Press, 1995.

Yeung, Y.M., and Y.W. Sung, eds. *Shanghai: Transformation and Modernization under China's Open Policy.* Hong Kong: Chinese University of Hong Kong, 1996.

Yin, J.Z. et al., eds. *Tizhi Gaige yu Shehui Zhuanxing* (System Reform and Social Transformation). Shanghai: Shanghai Social Science Press, 2001.

Yin, Q., and G. White. "The Marketization of Chinese Higher Education: A Critical Assessment." *Comparative Education* 30 (1994): 217–37.

Yuan, J.B., ed. *Xianggang Shilue* (Historical Sketch of Hong Kong). Hong Kong: Pillar Press, 1998.

Yuen, P.P. "The Sustainability of Hong Kong's Health Care Financing System: A Population Based Projection." *Policy Bulletin* 11 (May/June 1999): 5–9. Hong Kong: Hong Kong Policy Research Institute.

Zhang, H.M., D.T. Liu, and M.H. Gong. "Studies on Shanghai Housing Demand beyond 2000." *Chinese Real Estate Studies* 2 (2000): 108.

Zhang, X.H. et al., eds. *Shanghai Liudong Renkou de Xianzhaung yu Zhanwang* (Conditions and Prospect of Population Mobility in Shanghai). Shanghai: East China Normal University Press, 1998.

Zhang, Y.D., Z.L. Xia, C.M. He, and L.J. Sun. "Shequ Fuwu yu Renkou Laolinghua de Duice Yanjiu" (Community Services and Policy Response to Population Aging). In *21 Shiji Shangbanye Zhongguo Laoling Wenti Duice Yanjiu* (Research on China's Policy Response to Population Aging Issues in the First Half of the 21st Century), edited by W.F. Zhang et al., 321–52. Beijing: Hualing Chubanshe, 2000.

Zhang, Z.L. *Jindai Shanghai Chengshi Yanjiu* (Urban Studies on Modern Shanghai). Shanghai: Shanghai Renmin Chubanshe, 1990.

Zheng, G.C. *Lun Zhongguo Tese de Shehui Baozhang Daolu* (The Path of Social Security with Chinese Characteristics). Wuhan, China: Wuhan University Press, 1997.

Zhonggong Shanghai Shiwei Dangshi Yanjiushi, ed. *Shanghai Shehui Zhuyi Jianshe Wushinian* (Fifty Years of Socialist Construction in Shanghai). Shanghai: Shanghai Renmin Chubanshe, 1999.

Zhongguo Gongchandang Shanghai Dangshi Yanjiushi, ed. *Zhongguo Gongchandang zai Shanghai 1921–1991* (The Chinese Communist Party in Shanghai 1921–1991). Shanghai: Shanghai Renmin Chubanshe, 1991.

Zhongguo Minzheng Tongji Nianjian 1993 (China Civil Affairs Yearbook 1993).

Zhou, C.M. *Zhigong Yiliao Baoxian Zhidu Gaige Wenji* (Collected Articles on Employee Medical Insurance Systems Reform). Beijing: Employee Medical Insurance Systems Reform Project Team, Policy and Management Research Specialists Committee, Ministry of Health, 1995.

Zhou, N.H. "Strengthening the Connection Between Education and Economic Development: Major Issues in China's Educational Reform and Suggested Solutions." In *Social Change and Educational Development: Mainland China, Taiwan and Hong Kong,* edited by G.A. Postiglione and W.O. Lee. Hong Kong: Centre of Asian Studies, University of Hong Kong, 1995.

Zhou, N.Z., and F.P. Cheng. "Research on Higher Education in China." In *Higher Education Research at the Turn of the New Century: Structures, Issues and Trends,* edited by J. Sadlak and P. Altback. Paris: UNESCO Publishing, 1997.

Zhu, B.S. "*Shanghaishi di weilai liudong renkou zengchang he kongzhi*" (The Growth and Control of Population Mobility in Shanghai). Unpublished manuscript, 2000.

Zhu, R.J. Speech at the Opening of the First Session of the 9th National People's Congress of the People's Republic of China, 19 March 1998.

List of Editors and Contributors

David Chan is Associate Professor, Department of Applied Social Studies, City University of Hong Kong. His main research interests are comparative education policies and reforms in East-Southeast Asia. His publications include edited books, book chapters, and journal articles in *Comparative Education, Journal of Contemporary China, Asia Pacific Journal of Education, Chulalongkorn Educational Review,* and *Educational Policy Forum.*

Anthony Cheung is Professor, Department of Public and Social Administration, City University of Hong Kong. His main research interests include privatization and civil service and public sector reforms in Hong Kong and China. His latest books are *Public Sector Reform in Hong Kong: Into the 21st Century* and *Governance* and *Public Sector Reform in Asia: Paradigm Shift or Business as Usual?*

Alice Chong is Assistant Professor, Department of Applied Social Studies, City University of Hong Kong. Her research interests include long-term care and psychosocial therapy for older people, end-of-life issues (e.g., euthanasia, death, and breaking bad news to terminal cancer patients), consumer studies, and quality management in human services.

Gu Xingyuan is Professor, Institute of Public Health, Fudan University. He is Consultant to the Expert Committee on Policy and Management Research of the Ministry of Public Health. His major research interests are medical and health services and health insurance. He has published forty-one books and 204 articles.

Gui Shixun is Professor and Ph.D. supervisor, Institute of Population Studies, East China Normal University. He is President of Shanghai Population Studies Society. His research interests are population and social security for the elderly. He has published more than ten books, including *Sociology of Population,* and 150 articles.

Alex Kwan is Professor, Department of Applied Social Studies, City University of Hong Kong. He is the editor of *Capitalistic Welfare Development in Communist China—The Experience of Southern China and Aging Hong Kong—Issues Facing an Aging Society*. His research interests include social gerontology, aging welfare policies and services, and social work practice with older persons.

Kwok-yu Lau is Associate Professor, the Department of Public and Social Administration, City University of Hong Kong. He teaches courses on Housing Policy and Management in China and Social Policy. His research focuses on urban housing reform and privatization, public housing and economic change, and owners' involvement in property management.

Grace Lee is Associate Professor of Public and Social Administration at City University of Hong Kong. Dr. Lee's research areas are public policy and public management. Her most recent publication is *Reward for High Public Office: Asian and Pacific-Rim States*, Routledge, 2003 (edited with Christopher Hood and Guy Peters).

James Lee is Associate Professor, Department of Public and Social Administration, City University of Hong Kong. He specializes in teaching and research in housing policy and administration. He is a founding member of the Asian Pacific Network of Housing Research. His publications appeared in *Housing Studies, Policy and Politics*, and *Third World Planning Review*.

Simon Li completed his Master's degree in Public Policy and Public Administration at Concordia University, Canada, in 1993. Currently, he is a Ph.D. candidate of the Department of Social and Policy Sciences, University of Bath, United Kingdom. He is working on a thesis about Hong Kong employment and unemployment policy.

Ka-ho Mok is Associate Dean, Faculty of Humanities and Social Sciences and Associate Professor in the Department of Public and Social Administration, City University of Hong Kong. He has published extensively in the field of comparative education policy, and his most recent book is *Globalization and Educational Restructuring in the Asia Pacific Region*.

Raymond Ngan is Associate Professor, Department of Applied Social Studies, City University of Hong Kong. His research interests are social security reforms in China, Hong Kong, and Southeast Asia. His articles have appeared in *China Social Security, International Journal of Social Welfare, International Social Work, Journal of Aging and Social Policy*, and *Hallym International Journal of Aging*. He is currently preparing a monograph on Hong Kong's Mandatory Provident Funds.

King-lun Ngok teaches in the Department of Public and Social Administration, City University of Hong Kong. His major research interests include comparative social policy, labor law, and labor relations in China. His publications appear both in international and local journals including *Problems of Post Communism, Issues and Studies, Chinese Law and Government, Education and Society, Asian Profile,* and *Asian Review.*

Tang Anguo is Professor, Institute on Higher Education, East China Normal University. He is a member of the National Higher Education Establishment Council. His major research interests are higher education management and education system reform. He has published seven books and over sixty articles.

Wang Daben is Associate Professor, Institute of Population Studies, East China Normal University. He is a director of Shanghai Labor Studies Association. His major research interests are labor and employment and community development. He has published more than ten research reports and twenty articles.

Lynn White teaches in the Woodrow Wilson School, Politics Department, and East Asian Studies Program at Princeton University. He has written books about Shanghai and articles about Taiwan. His recent works include *Unstately Power: Local Causes of China's Economic Reforms* and *Unstately Power: Local Causes of China's Intellectual, Legal and Government Reforms.* His current project is a Taiwan-Jiangnan-South East Asian comparative book.

Linda Wong is Associate Professor, Department of Public and Social Administration, City University of Hong Kong. Her books include *Marginalization and Social Welfare in China, Social Change and Social Policy in Contemporary China,* and *The Market in Chinese Social Policy.* Her research interests include social welfare, urban migration, unemployment, and the non-state welfare sector in China.

Wu Duo is Professor, Institute of Sociology, East China Normal University. He is Deputy President of the Chinese Sociological Association. His research interests are urban sociology and community studies. He has published more than ten books, including *Sociology,* and more than one hundred articles.

Ray Yep is Assistant Professor in the Department of Public and Social Administration, City University of Hong Kong. His major research interests are state-business relations, political economy of rural reforms in China, and Chinese rural entrepreneurs. His publications have appeared in international and local journals, including *Pacific Affairs, Journal of Contemporary China, Political Quarterly,* and *Hong Kong Journal of Social Sciences.*

Ngai-ming Yip is Assistant Professor in the Department of Public and Social Administration, City University of Hong Kong. His research interests include social security policy and the social implications of housing and housing management. He has written on the social security system of Hong Kong, social capital, and neighborhood and housing management.

Zhang Yongyue is Professor, Ph.D. supervisor and Dean of Dongfang Institute of Real Estate, East China Normal University. He is Vice President of the Shanghai Real Estate Economics Association. His research interests are housing and real estate economics. He has published eleven books, including *New Real Estate Economics*, and over 100 articles.

Zhu Baoshu is Professor and Ph.D. supervisor, Institute of Population Studies, East China Normal University. He is a director of the China Population and Environmental Studies Association. His research interests are population geography and human resources development and management. He has published three books, including *Urban Extension and Labor Transition*, and more than fifty articles.

Index